TP13008352
SOS

Identifying Microbes by Mass Spectrometry Proteomics

Identifying Microbes by Mass Spectrometry Proteomics

Edited by
Charles H. Wick

CRC Press
Taylor & Francis Group
Boca Raton London New York

CRC Press is an imprint of the
Taylor & Francis Group, an **informa** business

CRC Press
Taylor & Francis Group
6000 Broken Sound Parkway NW, Suite 300
Boca Raton, FL 33487-2742

© 2014 by Taylor & Francis Group, LLC
CRC Press is an imprint of Taylor & Francis Group, an Informa business

No claim to original U.S. Government works

Printed on acid-free paper
Version Date: 20130621

International Standard Book Number-13: 978-1-4665-0494-3 (Hardback)

Library of Congress Cataloging-in-Publication Data

Identifying microbes by mass spectrometry proteomics / edited by Charles H. Wick.
    p. ; cm.
    Includes bibliographical references and index.
    ISBN 978-1-4665-0494-3 (hardcover : alk. paper)
    I. Wick, Charles Harold, editor of compilation.
    [DNLM: 1. Mass Spectrometry--methods. 2. Proteomics. QC 454.M3]

    QP519.9.M3
    543'.65--dc23   2013019472

Visit the Taylor & Francis Web site at
http://www.taylorandfrancis.com

and the CRC Press Web site at
http://www.crcpress.com

This work is dedicated to Professor Bjorn F. Hrutfiord, PhD

Contents

Preface

Writing a book on mass spectrometry methods for detecting and identifying micro-organisms took on the aspect of trying to summarize an entire epoch in a few words. Starting with the first instruments in the 1970s to the present, the discipline of mass spectrometry has evolved and has as many branches as an old oak tree. It became simpler when focused on one simple branch, which shows that microorganisms can be detected, identified, and classified using mass spectrometry, proteomics, appropriate software, and fast computers. The technology works fine. In fact, this process works better than "fine"; it has the ability to detect and classify those microbes that are not sequenced—those not in the genetic databases. Because of the exceptional standardization and organization of the genomic information available, it is a simple process to sort this information by using appropriate software and a computer. We have accomplished tasks in minutes using software which used to take weeks of wet laboratory work. All microbes can be classified according to their degree of match to their taxonomic hierarchy.

Microbe identification is actually very simple when you consider that all that is required is the detection of a few peptides unique to each microbe. Since peptides can be determined by mass spectrometry methods and these can in turn be sorted by software, the unique peptides of all the microbes can be determined. The process includes the following four steps: (1) all the available sequenced microbes are obtained from the national GenBank; (2) software, such as ABOid™, determines all the unique peptides for these microbes; (3) mass spectrometry provides the peptide information for all microbes in their samples; and (4) then back to ABOid™, which sorts out the unique peptides and determines the accurate identification and classification. This is important when considering that the detection of a few unique peptides out of thousands available for a microbe is easier than looking for a specific marker in a complex media. The result is accurate identification and classification over and over again not based on a few markers but based on the detection of a few out of thousands of unique markers (peptides).

This book is a snapshot in a rapidly growing, exciting area of science, and we can expect that updates will be the subject of a new work. It is clear that we are breaking new ground every day, and it will be fun to see where this approach takes us.

Simply detecting and identifying microbes using such an easy system is terrific. The future is exciting. We can expect miniaturized mass spectrometers in the not-so-distant future, and because the software can be ported to multiple platforms, this capability will be handheld and available to everyone. This capability changes everything. Imagine the fun of being able to go around and identify the microbes in the yard with a youngster. The possibilities are limitless.

MATLAB® is a registered trademark of The MathWorks, Inc. For product information, please contact:

The MathWorks, Inc.
3 Apple Hill Drive
Natick, MA 01760-2098 USA
Tel: 508-647-7000
Fax: 508-647-7001
E-mail: info@mathworks.com
Web: www.mathworks.com

Acknowledgments

I would like to express my gratitude to the many scholars who have shown the motivation and enthusiasm to develop the capabilities of mass spectrometry for more than 40 years. There are numerous scientists and engineers who have made these devices smaller, better, and more useful.

I would especially like to thank those who made this book possible: Patrick McCubbin, who labored to grind and process a wide variety of samples for analysis, for his humor and dedication, which were both welcome and rewarding; Dr. Rabih Jabbour and Dr. Jacek Dworzanski, who operated the mass spectrometers; and Alan Zulich, for his steadfast support. Particular appreciation is given to my son, Harrison Wick, for placing this manuscript into the final format.

Editor

Dr. Charles H. Wick, PhD, is a retired senior scientist from the U.S. Army Edgewood Chemical Biological Center (ECBC), where he has served both as a manager and research physical scientist and has made significant contributions to forensic science. Although his 40-year professional career has spanned both the public sector and the military, his better-known work in the area of forensic science has occurred in concert with the Department of Defense (DOD).

After earning four degrees from the University of Washington, Dr. Wick worked in the private sector (civilian occupations) for several years, leading to a patent, numerous publications, and international recognition among his colleagues.

In 1983, Dr. Wick joined the Vulnerability/Lethality Division of the United States Army Ballistic Research Laboratory, where he quickly achieved recognition as a manager and principal investigator. It was at this point that he made one of his first major contributions to forensic science and to the field of antiterrorism. His team was the first to utilize current technology to model sublethal chemical, biological, and nuclear agents. This achievement was beneficial to all areas of the DOD, as well as to the North Atlantic Treaty Organization (NATO), and gained Wick international acclaim as an authority on individual performance for operations conducted on a nuclear, biological, and chemical (NBC) battlefield.

During his career in the U.S. Army, Wick rose to the rank of lieutenant colonel in the Chemical Corps. He served as a unit commander for several rotations, a staff officer for six years (he was a division chemical staff officer for two rotations), and as a deputy program director of biological defense systems and retired from the position of commander of the 485th Chemical Battalion in April 1999.

Dr. Wick continued to work for the DOD as a civilian at ECBC. Two notable achievements, and one which earned him the Department of the Army Research and Development Award for Technical Excellence and a Federal Laboratory Consortium Technology Transfer Award in 2002, include his involvement in the invention of the integrated virus detection system (IVDS), a fast-acting, highly portable, user-friendly, extremely accurate, and efficient system for detecting the presence of, screening, identifying, and characterizing viruses. The IVDS can detect and identify the full spectrum of known, unknown, and mutated viruses, from AIDS to hoof and mouth disease, to West Nile virus, and beyond. This system is compact, portable, and does not rely upon elaborate chemistry. The second, and equally award winning, was his leadership in the invention of the method for detecting microbes using mass spectrometry proteomics. These projects represent determined ten-year efforts and a novel approach to the detection and classification of microbes from complex matrices.

Throughout his career, Dr. Wick has made lasting and important contributions to forensic science and to the field of antiterrorism. He holds several U.S. patents in the area of microbe detection and classification. He has written more than 45 civilian and military publications and has received myriad awards and citations, including the Department of the Army Meritorious Civilian Service Medal, the Department of the Army Superior Civilian Service Award, two United States Army Achievement Medals for Civilian Service, the Commander's Award for Civilian Service, the Technical Cooperation Achievement Award, and 25 other decorations and awards for military and community service.

Contributors

Dr. Samir V. Deshpande, DSc, is a senior bioinformatics scientist who, during his 20-year professional career, has designed software solutions to complex problems. His recent and most notable achievement has been the creation of the 100K line code of the successful program called ABOid™, which is credited with the discovery of virus and fungi associated with honeybees.

After earning an MS in electronics from Sardar Patel University, Dr. Deshpande earned distinction as a software engineer for Microsoft. He then joined the U.S. Army Edgewood Chemical Biological Center (ECBC) as a contractor with Science Technology Corporation. Thus began an outstanding career in developing bioinformatic methods for the analysis and classification of microbes. He then earned a DSc from Towson University and continued this career, becoming the leader and bioinformatics expert of the program that invented and discovered the means for detecting and identifying microbes using output from mass spectrometry.

Dr. Deshpande's experience and skill in software analysis along with his experience in design and development with tools like Visual Studio.NET, ASP.NET, Java, MATLAB®, LabView, Perl 5.2, PHP, XML, and C++ and a host of other methods provided the means to determine the complicated interrelationships among microbes. This unique ability resulted in the patented software platform known as ABOid™. This suite of programs allows for the first time a comprehensive means for detecting and classifying microbes to strain level using their unique peptides, a capability that has benefits to all areas of the Department of Defense, the study of infectious disease, and the discovery of new and emerging microbes.

Dr. Deshpande's research interests include the development of proteomic sequences data warehouse, proteomic identification algorithms, and bioinformatics application pipeline development using distributed computing. He has many peer-reviewed publications in reputed international journals. He has given numerous presentations and contributed to the paper that was awarded the Best in Basic Research by DTRA. In the field of data warehousing, data reduction, and archival, his research continues to involve design and creation using databases like Oracle 10g, MYSQL 5.2, and SQL Server 2005. His goal is to further advance this capability so that someday it will function on handheld devices and be useful to the general public.

Dr. Deshpande likes to teach data analysis and statistical software design and development. He enjoys teaching and has helped many students advance in their understanding of the important discipline of bioinformatics.

Dr. Michael F. Stanford, PhD, is a research physicist with more than 25 years of research and technical management experience within the DOD, NASA, and FAA as well as in academia and industry. After receiving his bachelor's degree in physics from Richard Stockton College, Pomona, New Jersey, in 1975, Dr. Stanford began his graduate studies in biophysics at Downstate Medical Center, State University of New York (SUNY), and received his doctorate in biophysics from the same university in 1982. He received his commission in 1983 as an O-3 (lieutenant) in the navy and served at the Naval Aviation Research Laboratory (NAMRL) in Pensacola, Florida. He served as a radiation specialist officer and concentrated his research on the detection and effects of ionizing and nonionizing radiation on biological systems. During his tenure at NAMRL, he represented the command at national and international conferences, presenting papers on the research in radiation effects ongoing at NAMRL. He retired from the Navy Reserve as an O-5 (commander) in 1997.

Upon completion of naval active duty in 1986, Dr. Stanford accepted a position with the BDM Corporation in Albuquerque, New Mexico, and served as a test and evaluation engineer. There, he participated in the development of the Corps Battle Analyzer (CORBAN) air and ground combat model. He also provided human factors expertise for calculating weapons' effectiveness and probability of kill determination. During this time, his work also focused on researching command, control, communication, countermeasures (C3CM) for the Joint Test Force (JTF) at Kirkland Air Force Base, New Mexico.

From 1988 through 2000, Dr. Stanford supported NASA at the Johnson Space Center in Houston, Texas. He participated in the development of a proton and heavy ion detector for assessing the radiation risk in low earth orbit. Additionally, he participated in the development of the tissue equivalent proportional counter (TEPC), which provided data on the secondary radiation field inside the space shuttle and space station (both these detectors are currently part of the space station and space shuttle suite of detectors). He was also involved in astronaut training on the nature of the space environment and its hazards.

In 2000, Dr. Stanford worked for Northrop Grumman Corporation in support of the Aviation Security Research and Development Laboratory at the FAA (after 9/11 TSA) William J. Hughes Technical Center, Atlantic City, New Jersey. He served as the program manager of the Trace Explosive Detector Program and provided technical expertise during the test and evaluation of technologies procured from multiple vendors.

In 2004, Dr. Stanford joined the Point Detection Team at Aberdeen Proving Ground (APG) and has been involved in the development of new technologies designed to detect and identify a host of biothreats on the battlefield and in the homeland.

Abbreviations

ABC	Ammonium bicarbonate
ABO	Agents of biological origin
ABOid™	Agents of biological origin identification
AHTS	Aerosol-to-hydrosol transfer stages
ANN	Artificial neural network
APC	Agent Containing Particles
APCI	Atmospheric pressure chemical ionization
AP/MALDI	Atmospheric pressure matrix-assisted laser desorption/ionization
ATCC	American Type Culture Collection
BAMS	Bioaerosol mass spectrometry
BEADS	Biodetection enabling analyte delivery system
BWA	Biological warfare agent
CAD	Collisionally activated dissociation
CI	Chemical ionization
CID	Collision induced dissociation
DART	Direct analysis in real time
DESI	Desorption electrospray ionization
DTT	Dithiothreitol
EAM	Energy-absorbing molecule
ECD	Electron capture dissociation
EI	Electron impact (ionization)
EI-MS	Electron impact ionization mass spectrometry
ELDI	Electrospray-assisted laser desorption/ionization
ELISA	Enzyme-linked immunosorbent assay
ESI	Electrospray ionization
ESI-FTICR	Electrospray ionization-Fourier transform ion cyclotron resonance
ETD	Electron transfer dissociation
FAME	Fatty acid methyl ester
FT	Fourier transform
FT-ICR	Fourier Transform-ion cyclotron resonance mass analyzer
GC	Gas chromatography
HCA	Hierarchical cluster analysis
HF-FIFFF	Hollow-fiber flow field-flow fractionation
HPLC	High-performance liquid chromatography
IMAC-Cu	Immobilized copper cations
IRMPD	Infrared multiphoton dissociation
IVDS	Integrated virus detection system
LIMBS	Laser-irradiated magnetic bead system
LIT	Linear ion trap
LIT/FTICR	Linear ion trap/Fourier transform ion cyclotron resonance analyzer

LV	Latent variable
MAB	Metastable atom bombardment
MAB-Py-MS	Metastable atom bombardment ionization pyrolysis mass spectrometry
MALDI	Matrix-assisted desorption/ionization
MS	Mass spectrometry
MS/MS	Tandem mass spectrometry
MSP	Mass spectrometry and proteomics
NCBI	National Center for Biotechnology Information
PCA	Principal component analysis
PCR	Polymerase chain reaction
Pfa	Probability of false alarms
PLS	Partial least square
PLS-DA	Partial least squares-discriminant analysis
PMF	Peptide mass fingerprinting
PTM	Posttranslational modification
Py	Pyrolysis
Q	Quadrupole
QIT	Quadrupole ion trap
Q/LIT/Q	Quadrupole/linear ion trap/quadrupole
QQ	Two quadrupoles
SASP	Small acid soluble protein
SAX2	Strong anion exchange
SEC	Size exclusion chromatography
SELDI	Surface-enhanced laser desorption ionization
SIM	Selected ion monitoring
SNP	Single nucleotide polymorphism
SRM	Selected reaction monitoring
sub-fmole	10-16-10-17 mole
TFA	Trifluoroacetic acid
TIGER	Triangulation identification for genetic evaluation of risk
TMAH	Tetramethylammonium hydroxide
TOF	Time of flight
TPP	Trans-proteomic pipeline
TQ	Triple quadrupole
VNTR	Variable number tandem repeat
WCX2	Weak cation exchange

1 Detection and Identification of Microbes Using Mass Spectrometry Proteomics

Charles H. Wick

CONTENTS

1.1 INTRODUCTION

Since they were first discovered scientists have wanted to be able to identify and classify the many different microbes, in particular bacteria, fungi, and viruses. It was an exciting time of discovery during the development of the disciplines of microbiology, mycology, and virology. The results were that in general an academic specialty was offered in the three fields. Sometimes, the bacteriology and virology were offered in the same academic area of microbiology. One reason for this early separation was their size. Fungi were large multi-micron to millimeter-sized organisms, bacteria were of 0.5–2.0 μm, and viruses were nanometer-sized; generally three orders of magnitude

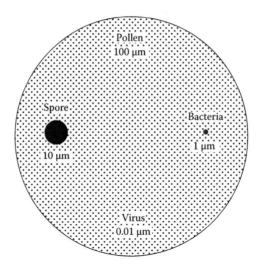

FIGURE 1.1 Relative size of microorganisms. Pollen, 100 μm; fungus, 10 μm; bacteria, 1 μm; virus, 0.01 μm.

separated the fungi, bacteria, and viruses (Figure 1.1). Many methods were developed for detection and identification of these microbes, and although some methods have many desired characteristics none of them satisfy all the criteria and none can detect and identify all three types of microbes in a single sample preparation.

The prospect of needing a rapid single detection platform presents many challenges. Some challenges are unique to epidemics and others are common for all testing situations (Klietmann and Ruoff 2001). Ideally, detection platforms should be capable of rapidly detecting and confirming hostile microbes, including modified or previously uncharacterized ones, directly from complex matrix samples, with no false results. Furthermore, the instrument should be portable, user-friendly, and capable of testing for multiple agents simultaneously.

1.2 BACTERIA

Early efforts to classify and identify bacteria concentrated on growing them, learning what they metabolized and other physical features, such as color, edge of colony characteristics, if they were round or rod-like, and whether they had an ability to stain with iodine and the like. These differences were used to classify them into groups for naming purposes. The importance of this ability was naturally directed at infectious diseases and the ability of professionals in the art to identify the same microorganism. Worldwide, this capability quickly helped in the control of infections and outbreaks of infections associated with historical epidemics. The discovery of antibiotics to control bacteria fueled a rapid increase in scientific work, and the discipline of microbiology grew along these lines of research.

The common and historical bacteriological methods have provided a basis for identification and frequently are still the routine identification methods. These techniques are generally based on the determination of the morphology, differential staining, and physiology of a bacterial isolate. These tests can be performed by means of miniaturized and automated substrates that utilize screening methods to classify/identify the isolate. One popular system in this category is the VITEK (BioMerieux, Hazelwood, MO) and MicroLog (BioLog, Hayward, CA) that utilize metabolizable substrates and carbon sources or susceptibility to antimicrobial agents. Such systems have been used to identify microbes such as *Bacillus anthracis* (Baillie et al. 1995), *Yersinia* spp. (Linde et al. 1999), and other pathogens (Odumeru et al. 1999). In addition, BioLog introduced a "dangerous pathogen" database to its MicroLog system. Although such techniques are still considered standard practice for the isolation and identification of infectious microbes, they are lengthy and it can take days for obtaining even preliminary results.

1.2.1 Immunoassay

To increase the specificity of detection methods, immunoassays gained popularity despite the fact that they can test for only one analyte per assay. This limitation means that multiple simultaneous or sequential assays must be performed to detect more than one analyte in a sample or specimen. Advances in assay design and in matrix format have resulted in development of multiplex assays that can be performed on multiple samples simultaneously by automated systems. However, the specificity of immunoassays is limited by antibody quality, and sensitivity (detection limits $\sim 10^5$ colony-forming units [cfu]) is typically lower than with polymerase chain reaction (PCR) and other DNA-based assays. With improvements in antibody quality and assay parameters, it may be possible to increase immunoassay sensitivity and specificity in the future. Many different immunoassay formats are currently commercially available for a wide variety of detection needs. Many formats are similar to, or derived from, the classic sandwich assay based on the enzyme-linked immunosorbent assay (ELISA) design (Murray et al. 2003).

1.2.2 Nucleic Acid–Based Methods

Nucleic acid–based methods usually combine PCR amplification with simultaneous detection of amplified products based on changes in reporter fluorescence. For specific detection, the change in fluorescence relies on the use of dual-labeled fluorogenic probes. An increase in fluorescence indicates that the probe has hybridized to the target DNA, and this principle is used for a variety of tests that rely on quantitative presence/absence of targeted sequences. However, the main PCR format used for biothreat agents is usually specific target detection, and a wide variety of primer and probe combinations are available from many companies in a multitude of configurations. Many of these specific target configurations rely on mechanistic variants in the primer/probe construction and combinations which can include TaqMan probes and other primers (Westin et al. 2001). They are available commercially and can be customized. Recently, several companies offer PCR kits in various formats for detection of biothreat agents. These kits eliminate the need for extensive primer/ probe design and facilitate rapid detection and monitoring programs. Many of the nucleic acid approaches for detection of biothreat agents are described in review articles (Ivnitski et al. 2003). Additional approaches also investigate methods to detect nucleic acids from target agents of interest using isothermal amplification or directly from samples without using an amplification step.

1.2.3 OTHER METHODS

Other methods have been developed and used and like most other technologies it is not uncommon to have generations of such instruments in a laboratory. A broad spectrum and rapid detection method has been indicated for some time, particularly for use in movable fixed sites or mobile and other field applications.

1.3 FUNGI

Fungi are frequently discernible from each other by eye and often could be seen growing in the yard. This made early separation groups by descriptive taxonomic means simple even in ancient times; some were good to eat and some are not. It was important to early man to be able to separate into groups, resulting in specialists. The discovery of the micron-sized fungi came along later as technology enabled the discipline to expand and these microbes were then included in the classification schemes. Mainly their outward characteristics were used to name the different fungal species. Those fungal species that affected people other than being toxic were generally few and their study was usually a specialized field within the overall medical microbiology discipline. However, those that affected plants, trees, and food sources were also important and the field of mycology evolved among these disciplines.

1.4 VIRUSES

The discovery of viruses resulted in the development of new methods and improvements in technology. Because of their small size, they were difficult to visualize and helped in the improvements in both scanning and transmission electron microscopes. The resulting discovery of the multitude of different surface features, sizes within the nanometer community, and other features resulted in their early characterization. Names of viruses were frequently associated with host and type of illness.

1.5 ADVANCEMENTS IN DETECTION

For many years, this was the status—three groups of scientists, three disciplines of research, and various naming schemes. This status all changed with the application of molecular biology and the ability to determine their genomic sequence. Standardization occurred and order was established among the microbes. This change swept through all three disciplines in a few years. Classification schemes were changed to move those microbes closely related genetically into the phylogenetic mapping scheme. All types of microbes were moved around within the old scheme to create a new classification. There were changes within the disciplines as a result, scientists relearned the new schemes and saw some of their microbes listed and published with new names along with the old names in parentheses.

1.5.1 GENOMIC SEQUENCING

The new genomic sequence phylogenetic mapping scheme for classification brought a standardization and new organization to the family, genera, species, and strains for all the microbes. This capability resulted in a new vigorous advancement in the detection and classification of all the microbes. It also enabled the sorting of microbes according to their bimolecular features, which enabled the new field of genomics and proteomics.

As science moved forward, more discoveries were made, new species of microbes discovered the capability to detect everything, and to keep track of all the different microbes became a challenge. Single probes became somewhat limited, multiplexed probes likewise found themselves trying to detect one more new microbe in the mix, and as emerging infectious agents have occurred this has continued to stress current methods. One new wrinkle in mix is the advent of the capability to manufacture new strains or even new microbes.

1.5.2 LIMITATIONS TO EXISTING METHODS

The current limitations of existing bioassays have become evident. They cannot generally identify unknown microbes. Identification is predicated on existing knowledge of a microbe. Specific antibodies or specific primer/probe pairs are required. Many attempts have been made to extend this capability, but it will always be short when confronted with the magnitude of the problem. Current techniques are expensive and frequently require expensive preparation, manufacturing, and reagents. Something else is indicated.

1.6 MASS SPECTROMETRY

Fortunately, during the last 25 years or more, mass spectrometry (MS) approaches to microbe identification have been investigated. Early devices were able to detect but not to identify microbes that paved the way to improvements which have resulted in MS techniques coupled with proteomic approaches that provide an acceptable next evolution in the detection and identification of microbes.

MS methods gather a wide range of information about microbes. The techniques are not limited by reagents or prior knowledge of a microbe. MSP is not a targeted approach to microbe identification.

Initially, MS-based methods used profiles of both pyrolytic products and fatty acids as specific microbial biomarkers for identification purposes. For example, the commercial microbial identification system (MIDI Inc., Newark, DE) continues to use gas chromatography (GC) of cellular fatty acid methyl esters for the identification of bacteria. This method has been used to identify and differentiate *Bacillus* spores and other potential biological warfare agents (BWAs). Furthermore, MIDI Inc. introduced the Sherlock Bioterrorism Library that can be added to its identification system to specifically target biothreat agents and challenge organisms. The MIDI Sherlock system containing the MIDI BIOTER database has been awarded AOAC Official Methods of Analysis status for confirmatory identification of *B. anthracis*

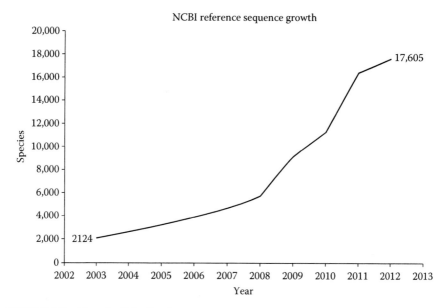

FIGURE 1.2   Growth of GenBank.

(AOAC International 2004). On the other side, the chemical biological (CB) MS system developed by DOD used a pyrolytic processing of biothreat agents to generate fatty acid methyl esters–dominated products analyzed by MS-based methods for discrimination of bioagents. Unfortunately, the specificity of these methods is relatively low.

Advancements in genomics, hardware, and software have made this new capability possible. A rapid increase in the number of sequenced microbes, bacteria, fungi, and viruses has occurred in the last few years and this trend is expected to continue (Figure 1.2).

MS have improved during the last several years, and in some cases what was dreamed possible only a few years ago is now taking place in our laboratories. Very fast acquisition (thousands of peptides in minutes) and high resolution (100,000 at mass-to-charge [m/z] ratio 400) have resulted in sensitivities (sub-femtomole = 10^{-16} to 10^{-17} mol by LC-MS) that are continuing to improve. Computer capabilities have improved with rapid acceleration during the last few years; it is now possible to make calculations that took longer than a week to perform a couple of years ago in less than a few minutes. All this capability is rapidly improving at such a rate that it is not unusual to see in an active laboratory several generations of various types of equipment used to evaluate microbes. The change is taking place as research is taking place and results from 1 year are frequently improved by new equipment in the next year or two. Frequently, what is limited by hardware or computers can see direct improvement in sensitivity and time to analyze in months.

The convergence of genomic sequencing, MS advancements, faster computers, and software have made it possible to sequence and analyze peptides using the MS. Software can be used to calculate all the known peptides for a sequenced microbe and then sort out all the microbes from each other to determine those peptides that are unique to a given bacterium, fungus, or virus. It is then simple to construct a phylogenic tree to identify an organism or organisms and relate them to their near neighbors. This is a genetically accurate and sensitive method to detect and identify all the microbes in a complex mixture.

It sounds simple enough: pick an MS, select a fast computer, and utilize appropriate software and you can suddenly detect and identify microbes. Well, scientists and engineers have been busy and there are many types of MS, several types of computers, and many sorts of software. Advancements historically have proceeded over a wide front. It seems that advancements have come at a fast pace for MS hardware, computers, and software.

MS proteomics (MSP) can detect and classify all three types of microbes. The technology has been thoroughly tested and many double-blind trials have proven its capability. It has high accuracy similar to common PCR means. The results also demonstrate that MSP can detect and identify microbes that are not sequenced (not in the genetic databases) to the level that can be matched based upon their genetic similarities, e.g., some *Escherichia coli* are sequenced and some are not. MSP detects and identifies a genus of *E. coli* for those not sequenced. All microbes can be classified according to the degree of match to their taxonomic hierarchy, making the detection and grouping of unknown microbes scientifically sound.

When MSP methods are combined with other methods such as the integrated virus detection system (IVDS), greater capabilities are realized. In the IVDS-MS method, many samples can quickly be analyzed for the presence of viruses and only the positive samples continue to the MS for identification. This combination also improves the sensitivity of detection by concentrating and separating the viruses from complex mixtures. Such hybrids improve discrimination of negative and positive samples and improve overall identification time.

It is appropriate then to have a look at the different types of MS that have been developed and their operating principles, and to briefly examine how they are used. Since these are different types of MS, it is not surprising that they have in general different preparation methods. The analysis of the MS files is then determined by what sort of information can be obtained.

1.6.1 A Single Biodetection Method Based on MSP

It is clear that a combined microbe detection platform must be capable of detecting a variety of microbes in complex samples. This multiplex capability is vital because samples may contain toxins, bacteria, viruses, or other types of analytes. In some instances, microbes may have been deliberately altered through genetic, antigenic, or chemical modifications or may represent new or uncommon variants of known microorganisms. Such modifications can make detection difficult using the common methods. Therefore, the only way to overcome these problems in a timely fashion would be rapid sequencing of nucleic acids or to deduce nucleotide sequence information from amino acid sequences of proteins. The latter approach has the advantage of including protein toxins, which can be easily modified to escape detection by any method that does not rely on amino acid sequence information.

Developments in the area of MS allow for the application of this analytical platform for analysis of nucleic acids and proteins and obtaining sequence-based information about microbial world. This information is suitable for the detection, classification, and reliable identification of all microbes in a timely manner. Three important considerations are the sensitivity, the specificity, and the reproducibility of such a platform.

1.6.1.1 Sensitivity

An important consideration in biodetection is the collection and handling of samples. Airborne and waterborne samples must be generally concentrated from large volumes to detect low levels of target analytes. In many cases, airborne samples must also be transferred to a liquid because most detection platforms process only liquid samples. The efficiency of recovery from concentration and extraction procedures can vary and affects detection limits. Nevertheless, it is advantageous to isolate or concentrate target analytes prior to analysis in complex sample matrices such as powders or food.

In general, nucleic acid–based detection systems that use PCR followed by amplicon detection are more sensitive than antibody-based or MS-based methods that perform analysis directly from the sample, that is, without using any target amplification approach. For example, a PCR assay can detect 10 or fewer microorganisms in a short period of time (Bell et al. 2002; Ibekwe and Grieve 2003). However, PCR requires a clean sample and is unable to detect protein toxins and other non-nucleic acid–containing analytes such as prions. Furthermore, cultures of the target organism are not available for archiving and additional tests after PCR analysis. MS-based techniques targeting specific microbial constituents are less sensitive than PCR methods; however, MS can also be used for detection of PCR products. Hence, the sensitivity and specificity obtained with such MS-based technology can surpass any traditional method.

1.6.1.2 Specificity

Specificity is as important as sensitivity in the detection of biothreat agents. High specificity is important to minimize background signals and false-positive results from samples that are often complex, uncharacterized mixtures of organic and inorganic materials. Specificity can be affected not only by background particles, but also by high concentrations of competing antigens and DNA. In the case of PCR, its high sensitivity can also be a major weakness because contaminating or carry-over DNA can be amplified, resulting in false-positive results. MS-based methods that use sequencing information for detection and identification purposes are usually characterized by high specificity that is limited only by the sequencing information available in databases.

1.6.1.3 Reproducibility

In addition to sensitivity and specificity, reproducibility is an important requirement for detection platforms because systems that do not provide reproducible results are unreliable and may exacerbate a terrorist event. Many factors can affect the repeatability of bioassays, including the stability and consistency of reagents and differences in assay conditions. These variations can often be reduced by standardizing assay conditions and procedures; however, the best solution would be the use of reagentless assays that offer MS.

1.6.2 VERSATILE ONE-METHOD APPROACH TO DETECTION AND IDENTIFICATION

MS-based methods represent a broad group of highly versatile approaches that use precise mass measurements to infer identity of diverse biomolecules. Although for many decades the scope of investigated molecules was limited by their molecular mass and polarity, recent developments in soft ionization techniques like electrospray ionization (ESI), matrix-assisted laser desorption ionization (MALDI), or desorption electrospray ionization (DESI) substantially broaden the range of investigated species. Therefore, nowadays not only proteins and nucleic acids but also multimolecular complexes, and even whole viruses, can be mass analyzed by modern MS instruments and used to infer genetic information encoded in nucleotide and amino acid sequences. In this report, a group of highly versatile methods for the detection, identification, and classification of microorganisms are described, which are based on capabilities of modern MS techniques.

1.7 ABOUT THE CHAPTERS

The chapters are presented to tell the MS story in a way that helps the reader gain a general understanding of the many variables associated with using MS methods for microbe identification. With this in mind, Chapter 2 starts with a review of the various mass analyzers, followed by Chapter 3, which discusses the various methods used to analyze MS files. Building on this foundation, it then follows that the methods associated with genomic approaches and MS are discussed in Chapter 4, followed by sample collection and processing in Chapter 5. It is here that the reader is taken to the next step in the evolution of MS and microbe detection by considering computer software and bioinformatics approaches in Chapter 6. Three applications of the software/bioinformatics approach are given in Chapter 7. These three examples demonstrate the versatility of MSP in the detection, identification, and classification of microbes. Chapter 8 is a survey of commercially available MS-based platforms currently suitable for microbe detection using the MSP approach. Chapter 9 is a discussion of future trends.

MS techniques determine the molecular mass of compounds by separating ions according to their m/z ratio; therefore, any species to be analyzed by MS has to be ionized in the first place. The ionization methods frequently used for analysis of microbial constituents are described in Chapter 2. Ions formed in the ion source of the MS are next moved to the analyzer section that uses combinations of electric and magnetic fields to separate and detect ions according to their m/z values. Furthermore, molecular ions formed in the ion source may be additionally excited and forced to dissociate into products that are then immediately mass analyzed. The details of the most popular mass analyzers and their hybrids, which are used for microbial detection and identification purposes, are continued in Chapter 2.

The collection of environmental samples for microbial analysis represents a crucial step on the pathway to reliable results; therefore, diverse processing methods are reviewed in detail in Chapter 3. However, the raw results of MS analysis of microbial species usually represent huge amounts of data that have to be transformed into meaningful information. The chemometrics and bioinformatics (software) approaches for data mining and discrimination of microbes are presented in Chapter 6.

In recent literature, the main MS-based approaches used for the characterization, classification, and identification of microorganisms can be grouped in two broad classes. The first class of methods relies on mass spectral library building followed by fingerprint matching of biomarkers, or by using pattern recognition techniques of unknown products, which correlate with the presence of a particular agent. These methods are useful for microorganisms for which complete genomic/proteomic databases are not available. Pattern recognition-based approaches using different analytical techniques such as pyrolysis products, chemical ionization (CI) of fatty acid methyl esters, and MALDI-MS are described in Chapter 6, with a special emphasis on MALDI applications due to its ability to produce information-rich mass spectra in a mass range of up to several tens of kDa.

In Chapter 4, sequence-based methods for the classification and identification of microbial agents are discussed. These methods employ accurate mass measurements and/or tandem MS (MS/MS) techniques to infer base compositions of PCR amplicons or amino acid sequences of microbial proteins. However, related methods based on the comparison between observed protein masses in the MALDI spectrum and protein databases are also covered.

Chapter 5 details a few of the specific sample preparation techniques, and Chapter 6 gives an explanation on how the ABOid™ software processes MSP files to given microbe detection and identification results.

Chapter 7 details several examples of the application of MSP methods to detect, identify, and classify microbes. These include double-blind studies, multiple microbes in a single sample. How to evaluate differences within a single microbe as in the Pandemic H1N1 2009 is demonstrated. The successful detection of microbes from complex environmental samples came from honeybees.

In Chapter 8, a survey of commercially available MS-based platforms that could be used for microbial detection and identification is presented, while in Chapter 9, future trends of using MS techniques for the detection and identification of biological agents are outlined.

2 Mass Analyzers and MS/MS Methods for Microbial Detection and Identification

Michael F. Stanford

CONTENTS

2.1   INTRODUCTION

Since the 1970s, analysis by MS methods has exploded into a wide and extensive network of scientific disciplines. The early vanguards were chemists, biochemists, biologists, and then molecular biologists; up to now it is difficult to say who is in the lead. MSs are common in a modern laboratory and nearly all academic areas have their MS. It is now common to see a MS center in major universities that serve as wide range of academic disciplines. This has proved to be a viable solution as the state-of-the-art instruments continue to advance rapidly and it is not unusual to see substantial improvements in capability in months rather than years. An entering graduate student can expect to start mass spectra analysis on one machine and finish on a more sensitive, faster, and more capable MS.

Generations of MS in a single laboratory have created follow-on applications of this technology to other areas, and as operation and maintenance of these instruments continue to be simplified these new capabilities have possibilities. It will be presented in more depth later, but one of these possibilities is the capability of detecting microbes with less than a state-of-the-art MS. Advancements in software and computer speed have enabled faster analysis of MS files and often these files contain sufficient information for the classification of microbes.

Nevertheless, the first step is to examine the different types of MSs. Scientists and engineers have been busy and the result over the last 40 years is a variety of approaches to analysis by MS. This chapter will examine the major types of instruments and their various combinations used in microbe analysis.

Methods for sample preparation, instrument operation, analysis, and reporting of results are as varied as the types of platforms. It is appropriate to review those methods associated with the different MS.

Because of this variability, it is important to include the sample preparation methods according to MS techniques. It needs pointing out that, while it is possible to prepare a sample in less than 10 min for inserting into the instrument, sample preparation time may be reflected in the dynamic range of the instrument. Fast sample preparation does not always indicate the best results. Recent improvements in dynamic range, however, by several orders of magnitude indicate that additional increases in the dynamic range can be expected for MS methods, thus making short sample preparation times more attractive. Although not always indicated, it is assumed that sample preparation methods can be automated.

Instrument operation is a function of existing hardware and possible future improvements can be expected. The current descriptions are generic in nature and as the many types of MS systems, the demonstration of some small-sized units, and development of mini-MS systems are ongoing concerns, this would indicate that the size and weight of future MS systems will be suitable for mobile operations. Hardening and manufacturing to military specifications should likewise be a straightforward process. The trades between results, weight, power, and other requirements can be evaluated to create a future MS system that is optimized for specific applications, e.g., medical, emergency response, home use, and environmental applications.

Although mass spectral analysis is the process that has the greatest possibility for rapid improvement, it is the bioinformatic tools that have made it possible to quickly process large MS files in a short time. Such software improvements over the last few years have reduced processing time from hours to seconds. Similar improvements in hardware and software can be expected and the additional speed will further reduce mass spectral data processing time.

The following types of MS are thus presented to simply demonstrate the various approaches to MS analysis and the creative approach that scientists and engineers have taken over a short time.

MS techniques determine the molecular mass of compounds by separating ions according to their *m/z* ratio; therefore, any species to be analyzed by MS has to be ionized in the first place. The ionization methods frequently used for analysis of microbial constituents are described. Ions formed in the ion source of the MS are next moved to the analyzer section that uses combinations of electric and magnetic fields to separate and detect ions according to their *m/z* values. Furthermore, the molecular ions formed in the ion source may be additionally excited and forced to dissociate into products which are then immediately mass analyzed. The most popular mass analyzers and their hybrids, which are used for microbial detection and identification purposes, are presented.

This chapter has two main sections: Section 2.2 is about ionization methods for microbial characterization and Section 2.3 is about the types of mass analyzers. This division is the result of the many approaches to microbial detection. Remembering that MSs were first in the preview of chemists, many of the approaches were from this line of thinking. In any case, all the MS approaches start with ionization. The evolution of the many types of mass analyzers and their different attributes is important as it provides choices in the future.

2.2   IONIZATION METHODS

Microbial agents are usually collected and undergo processing designed to obtain cellular constituents in a form suitable for ionization. The ionized species are mass analyzed and the data interpreted using computer-assisted methods. The complete analytical process is shown in Figure 2.1, which emphasizes the critical role of the ionization process for every application.

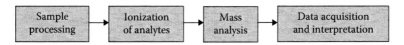

FIGURE 2.1   Schematic representation of microbial sample analysis by MS.

TABLE 2.1

Ionization Techniques Used for Analysis of Biological Compounds

Ionization Mode	Type of Analytes	Examples of Microbial Biomarkers
Electron impact (EI)	Low molecular mass, low polarity	Simple lipids and carbohydrates, e.g., fatty acids and sugars, fermentation products, cell pyrolyzates
Chemical ionization (CI)	Low to higher molecular mass, high proton affinity	Simple and complex lipids and cellular carbohydrates, pyrolyzates
Atmospheric pressure chemical ionization (AP-CI)	Low to higher molecular masses	Same as CI with the addition of proteins, DNA, and large biomolecules
Metastable atom bombardment (MAB)	Low to higher molecular mass, low and high polarity compounds	Various cellular components, e.g., pyrolyzates of cellular compounds
Electrospray ionization (ESI)	Low to high molecular masses, negative and positively charged molecules	Proteins, DNA, peptides, carbohydrates, intact and fragmented biomolecules
Matrix-assisted laser desorption ionization (MALDI)	Low to high molecular masses	Proteins, DNA, peptides, carbohydrates, intact and fragmented biomolecules; certain intact viruses and bacteria
Desorption electrospray ionization (DESI)	Low molecular mass	Lipids, carbohydrates, peptides

The application of MS to any class of species is challenging because gas-phase ions are required for analysis. Ionization processes, which are used for analysis of microbial constituents, are shown in Table 2.1 and are described briefly in this chapter.

2.2.1 ELECTRON IMPACT IONIZATION

The first ionization techniques that were applied to analysis of compounds derived from microbial agents were electron impact (EI) ionization and CI. Both of these techniques use a two-step process that consists of heat vaporization, followed by ionization, which occurs once the analyte is in the gas phase. The vaporization step limits the range of compounds that can be efficiently ionized before undergoing thermochemical transformations. These transformations are associated with the breakdown of covalent bonds and formation of products composed of lower-molecular-mass species and highly cross-linked polymeric residues. Hence, some researchers utilized thermal transformations of polar compounds under the influence of heat (pyrolysis) as a fast sample-processing procedure for the analysis of microorganisms through the characterization of volatile pyrolysis products that are easily ionized under EI and CI conditions.

2.2.1.1 Mechanism of EI Ionization

EI ionization is the result of interactions between analytes and a high-energy beam of electrons (usually, 70 eV) that are produced from a hot filament. The ionization

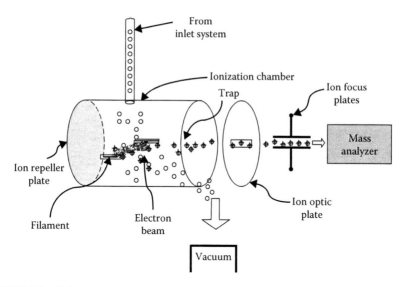

FIGURE 2.2   Schematic representation of EI ionization process.

occurs in a vacuum (10^{-5} to 10^{-4} torr) that is needed to minimize ion/molecule collisions. The source of electrons in EI ion sources is a metal filament that is electrically heated to incandescence temperature at which it emits free electrons. The emitted electrons are attracted to an anode (trap plate) situated on the opposite side of the ionization chamber from the filament/cathode (Figure 2.2). Under these conditions, collisions result in the expulsion of an electron from an analyte molecule (M) to form an electron-deficient radical cation, the molecular ion ($M^{+\bullet}$). The resulting molecular ions exhibit a distribution in their internal energies ranging from 2 to 8 eV. The radical cations with internal energies at the lower end reach the detector as intact ions without any further dissociation; however, ions with higher internal energy dissipate high internal energies through the dissociation of covalent bonds. This results in the ejection of a radical or neutral species to yield various ion products. These species may undergo further fragmentation if sufficient internal energy is present and each ion may display several modes of decomposition. Separation based on m/z ratio and measurement of the abundance of each ion gives the mass spectrum. Such fragmentation pattern may be used to elucidate the structure of the original molecule.

Attomolar sensitivities have been reported for MS analyses using the EI mode of operation. Therefore, EI ionization-based instruments found wide application in structural studies by providing spectra of fragment ions that can be traced to original structures. Moreover, the broad use of this ionization technique was substantially increased due to the availability of computer searchable databases of EI spectra for almost a million of organic molecules. In addition, there is extensive literature available that deals with the EI-MS analysis and specialized libraries of EI spectra for different classes of organic compounds that can be used for the identification of microbial constituents and for chemotaxonomic purposes.

2.2.1.2 Application of EI-MS for Detection and Identification of Microbial Biomarkers

EI-MS was broadly used in investigations of (a) volatiles from culture media, (b) various microbial metabolites, (c) pyrolysis products of whole cells and selected cellular constituents, and (d) in the selection of potential bacterial biomarkers for chemotaxonomic studies. Among bacterial constituents, lipids were most intensely studied using EI fragmentation spectra. For example, a commercially available MIDI system uses GC separation of methylated fatty acids (FAMEs) derived from microbial cells coupled with EI-MS analysis to determine specific FAME profiles. Such profiles are searched against a library containing thousands of microbial fatty acid profiles to establish identification. In addition, advancements of the GC-MS technique with EI as an ionization source provided reliable tools to investigate unique bacterial components as biomarkers for trace detection of bacteria in various matrices. For example, Fox used monitoring of selected ions generated under EI ionization of muramic acid, which served as a biomarker of bacterial peptidoglycan, for highly sensitive and selective monitoring of bacteria in environmental samples (Fox and Fox 1991). The same research group also used sugar profiles for discrimination between *B. cereus* and *Bacillus anthracis* strains based on the presence/absence of specific sugars identified using specific fragmentation patterns revealed during EI ionization of the GC effluents (Fox et al. 1993).

2.2.2 CHEMICAL IONIZATION

The CI approach is a softer ionization method in comparison to EI. CI is a result of the collision of an analyte with proton-rich reagent ions without elastic transfer of internal energy to the analyte molecules. Such chemical reaction results in the formation of protonated adduct ions, which represent the protonated molecular species of the analyte. Because much less energy is transferred to the sample molecule in converting it to an ion, less ion fragmentation occurs than in the case of EI ionization. CI can be performed in the EI ion source after additional supply of the reagent gas. The reagent gas is usually a small molecule, such as water, methane, isobutene, ammonia, or ethanol, that is introduced into the ion source to achieve a thousand times higher pressure in comparison to EI operation. When the sample enters the CI ion source, it reacts with the reagent gas ions to form new ions.

Quantitatively, the reagent gas ions concentration is kept higher than that of analyte molecules in order to ensure the efficient ionization of analyte molecules. The source of electrons in the CI source is a heated filament, which emits energetic electrons (100 eV) needed to penetrate the pressurized gas chamber. The reagent ions are usually used as protonated molecular ions of the reagent gas. For example, in the case of methane used as a reagent gas, protonated reagent ions CH_5^+ are formed, which react with analytes.

2.2.2.1 Positive Ions

Positive ion formation is the result of the proton transfer reaction between reagent gas cations and analyte molecules that usually produces intense protonated molecular ions (MH⁺). These molecular ions are even-electron species with most of the energy transferred to the protonated molecule. The proton affinity is usually governed by the analyte affinity and the type of reagent gas used during CI. It is important to notice that the ion/molecule reaction must be exothermic for protonation to occur in CI and thus the proton affinity of the analyte should be greater than the proton affinity of the reagent gas (Gorman et al. 1992; Yang et al. 1996).

2.2.2.2 Negative Ions

The formation of negative molecular ions during CI is the result of the reaction between analyzed molecules and negatively charged reagent ions such as chlorides and involves a direct electron transfer. Such electron transfer between analyte molecules and reagent gas ions is achieved when the thermal energy of the electron is around 0.1 eV. Negative ion CI is distinctively sensitive relative to positive ion CI due to the low-energy electrons that are transferred to analyzed molecules. This sensitivity is observed for compounds with high electron affinity such as halogenated hydrocarbons (Hunt et al. 1978).

2.2.2.3 Atmospheric Pressure CI

Atmospheric pressure CI (AP-CI) occurs by the low-energy electrons generated from a radioactive beta source or corona discharge where the electrons ionize the reagent gas and the ionized gas interacts with the analyte molecules (Bruins 1991). The AP-CI source is purged with nitrogen gas as a carrier and because of the relatively short path that analyte molecules have to pass such purging improves collisions between the analyte molecules and reagent ions. The presence of water will help in capturing the free protons whenever there is no stronger proton-affinity molecule than that of water present. As more basic molecules than that of water enters the AP-CI source, protons are transferred to it and collisions with reagent gas result in the formation of clusters of protonated species. AP-CI enhances sensitivity; however, it requires high purity sample and solvent. Moreover, AP-CI could be interfaced with separation instrumentations (i.e., LC) and capillary electrophoresis to enhance resolution of the analyte mixture (Bruins 1991).

2.2.3 Metastable Atom Bombardment

Metastable atom bombardment (MAB) is a relatively novel ionization method for MS analysis of small molecules (Bertrand et al. 2000). MAB has several advantages relative to electron and CI because it offers (a) excellent ionization efficiency, (b) highly reproducible fragmentation pattern (due to narrow energy distribution), and (c) user-selectable ionization energy. In MAB ionization, the excitation of gas molecules is achieved by corona discharge; however, the ions formed (M^+) during this process are deflected from the gas stream, while metastable atoms (M^*), that is, neutral atoms with internal energy elevated in relation to the ground state, are directed to an ion source and ionized organic molecules. Typically, metastables are formed from noble gases or nitrogen, and ionization reactions of types I and II (see the following reaction) occur between organic compounds (BC) and metastable species (A^*) in the collision cell:

$$A^* + BC \xrightarrow{\ \ I\ \ } A + BC^{+\bullet} + e^- \xrightarrow{\ \ II\ \ } \left[A + B^{+\bullet} + C + e^- \text{ or } A + B^+ + C^\bullet + e^- \right]$$

These reactions can occur, provided the excitation energy of the metastable atom is higher than the ionization potential of the compound. A reaction of type I involves an electrophilic reaction of the metastable A* with the analyte BC, resulting in a return of the metastable atom to its ground state (A) and ionization of the analyte as an odd-electron BC$^{+\bullet}$ species. More precisely, the molecular orbital of BC transfers one electron to a vacant orbital of the excited atom A*, leading to a simultaneous ejection of one electron from the outer shell of A*. When the amount of excitation energy transferred to the analyte is high enough, one or more bonds may break within BC$^{+\bullet}$, as indicated in reaction II.

Since the internal energy states available to metastables are constrained by quantum mechanics, all excess energy must be imparted to the analyte. Therefore, metastable ionization leads to a fragmentation pattern characteristic of the bond locations and their strengths, and the resulting fragmentation pattern is essentially independent of instrumental conditions. The amount of energy transferred in MAB is adjusted for each application by using gases in which the available metastable states differ in the amount of energy above their ground state. Interacting with low-energy metastables such as Xe*, Kr*, or N_2*, a typical analyte's molecular fragmentation is minimal. In short, a greater proportion of analyte spectral information is represented in intact molecular ions.

This ionization mode was applied for analyzing thermal fragments generated during pyrolysis (Py) of bacteria coupled with Py-MS measurements of bacterial cell components. For bacterial characterization by MAB, there is a choice between the available energetic extremes of He* (19.8 eV) and Xe* (8.3 eV). One selects a metastable species capable of ionizing biomarkers characteristic of cell identity but having too little energy to ionize background components. Generally, MAB gives greater spectral reproducibility than conventional 70 eV EI, and these capabilities commend Py-MAB-MS for the analysis of biological samples, sometimes with minimal preparation (Dumas et al. 2002).

2.2.4   ESI

ESI takes place by fine spraying a liquid solution of analytes into a strong electric field (2–4 kV) imposed between the capillary tube and the counter electrode of the MS inlet (Figure 2.3). Such a process produces clusters of charged droplets that consist of analyte with many solvent molecules. Protons are the charge carriers for the analyte–solvent clusters when ESI is performed in positive mode. The initial droplet size could vary from few to 50–60 μm in diameter and depends on the diameter of a nozzle and the electrospray flow rate. These droplets decrease in size due to the evaporation of solvent molecules and as the droplet diameter decreases, the charge density increases and repulsive forces promote electrohydrodynamic disintegration into many smaller droplets. The asymmetry in the charge distribution at the droplet surface will induce columbic explosion of the droplet, at which the repulsive forces among like charges overcome the cohesive forces of the solvent. In negative ion ESI, ion emission is accompanied by electron emission from the emitter or counter electrode. In order to avoid the loss of analyte due to a large-sized droplet, an assisted spray version is used where a sheath gas is introduced from the high-pressure side of the MS sample introduction area to the electrospray needle. The sheath gas, usually N_2, serves as drying stream in order to enhance evaporation and efficient removal of the solvent/water clusters from the charged analyte. The resulting analyte

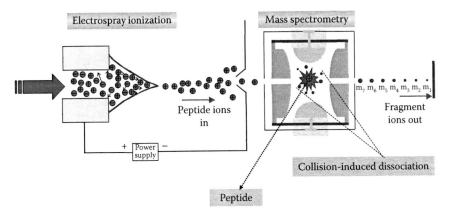

FIGURE 2.3   ESI source and its interface to a quadruple ion trap mass analyzer.

ions enter the MS through electrostatic lenses. One of the most useful attributes of ESI is its ability to interface MS with a popular separation technique such as HPLC and capillary electrophoresis, thus providing an efficient and highly sensitive platform for analysis of biomolecules presented in complex matrices.

The most extensively used mass analyzers for ESI are quadrupoles, time-of-flight (TOF) analyzers, and ion traps, which can handle the high pressure associated with the ESI interface.

2.2.4.1   Nano-ESI Source

The ionization efficiency of ESI depends on the presence of salts, buffers, and other additives or matrix components. Therefore, separation techniques are usually used prior to ESI-MS analysis (Bothner et al. 1995). However, many other approaches aimed to increase the sensitivity of MS analysis of ESI analytes were addressed (Gale et al. 1993; Wilm et al. 1994; Wilm and Mann 1996). The common feature of these studies is the use of narrow spray capillaries with nozzle diameters in the micrometer range. Such nano-ESI sources are used with flow rates below 1 µL/min and produce nano-droplets in the order of 100 nm in diameter. This approach offers several advantages in comparison to classical sources (Gabelica et al. 2002) like negligible consumption of solvents and high sensitivity (Schneider et al. 2003; Liu et al. 2004; Valaskovic et al. 2006).

2.2.4.2   Applications of ESI

ESI has inherent analytical advantages that allow it to be utilized for the analysis of different biological problems. Since the limitation of molecular mass is minimal, relatively large biomolecules have been successfully mass analyzed using ESI techniques that include even intact viruses or their chromosomes (e.g., coliphage T_4 DNA with nominal molecular mass of 1.1×10^8 Da) (Smith et al. 1994; Chen et al. 1995). Moreover, microbial carbohydrates, lipids, single-stranded DNA, RNA, proteins, and peptides were studied through ESI-MS and have been used for the detection, identification, and classification of microbes (Chenna and Iden 1993; McCluckey et al. 1995; Smith et al. 1995; Wickman et al. 1998; Arnold and Reilly 1999; Chen et al. 2001; Zhou et al. 2001; Dworzanski et al. 2004, 2006), and their toxins (Hua et al. 1993).

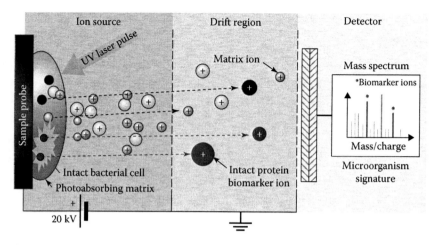

FIGURE 2.4 Principle of operation of MALDI-TOF-MS for microorganism identification. Intact biomarker ions are desorbed into the vacuum as a result of laser photon–matrix molecule interactions. All ions acquire kinetic energy (KE) proportional to their charge z. Ions with the same charge but different masses have different TOF through the drift region of the instrument. Calibration procedures correlate an ion's TOF, measured in an experiment, with its mass m. (From Ecelberger, S.A. et al., *Johns Hopkins APL Tech. Digest*, 25(1), 14, 2004.)

2.2.5 MALDI

MALDI is a method allowing for the ionization and transfer of a sample from the solid phase into gas phase that was introduced by Karas and Hillenkamp in 1988 (Karas and Hillenkamp 1988). This type of laser desorption is a soft ionization process achieved by bombarding a mixture of analytes and matrix with laser beam photons, which induce sample desorption and ionization. MALDI is a popular ionization technique that is most frequently used in combination with TOF-MS (Figure 2.4).

The desorption ionization process is usually performed in the vacuum chamber of these instruments (see Section 3.2.1); however, many other types of analyzers can be used for this purpose. The main advantage of the method is that it directly measures molecular masses at a very high speed (about 100 µs) and usually requires minimal sample preparation. Therefore, MALDI is extensively used as a powerful analytical tool for analysis of carbohydrates, lipids, and other thermally labile biomolecules like peptides, proteins, or nucleic acids (Hillenkamp et al. 1991; Chait and Kent 1992; Cotter 1992). MALDI usually produces single-charged molecules and thus resembles mass spectra produced by CI for low molecular compounds.

2.2.5.1 Matrices Used in the MALDI

Matrix and sample solutions are mixed prior to laser exposure, spotted on a solid metallic surface, and allowed to dry before submitting them to MALDI. The most commonly used matrices, which are used in MALDI analyses, are (a) α-cyano-4-hydroxycinnamic acid (CHCA); (b) 2,5-dihydroxybenzoic acid; (c) sinapinic acid; and (d) 3-amino-4-hydroxybenzoic acid. These matrices are characterized by high absorptivity of the laser radiation and their capability of forming fine crystalline solids during sample/matrix drying. In general, the more fine-grained and homogeneous the morphology of crystals formed with the analyte/matrix mixture, the more intense is the MALDI mass spectrum of the analyte.

It is essential for successful analysis by MALDI-MS to have suitable sample preparation conditions (Vorm and Roepstorff 1994; Cohen et al. 1997) because the quality of the matrix crystal formation depends on the matrix and solvent used and affects the sensitivity of analysis (Zhang et al. 1999). Although MALDI is more tolerant to the presence of buffers and salts than ESI, higher concentrations of these additives may adversely suppress the ionization by affecting the matrix crystal formation and the important interactions of the sample molecules with matrix crystals.

2.2.5.2 Applications of MALDI-TOF-MS

To date, MALDI-TOF-MS has been successfully used for the analysis of a wide range of different biomolecules that include peptides and proteins, oligosaccharides (Hsu et al. 2006; Wuherer and Deelder 2006; Fernandez et al. 2007), oligonucleotides (Li et al. 2006; Sun and Guo 2006), lipids (Vanrobaeys et al. 2005; Zambonin et al. 2006), and molecular aggregates with non-covalent interactions (Bolbach 2005; Yanes et al. 2007). The application of MALDI for the detection and identification of microorganisms is reviewed and extensively discussed in Chapter 6.

2.2.5.3 AP-MALDI System

The AP-MALDI technology was invented in 2000 (Laiko et al. 2000). This new sample ionization technique was coupled with many types of mass analyzers, including commercial ion trap-MS. The detection limit of the novel AP-MALDI-ion trap is 10–50 fmol of analyte deposited on the target surface for a four-component mixture of peptides with 800–1700 molecular mass in the range of 800–1700 Da. The possibility of peptide structural analysis during MS/MS and MS^3 experiments for AP-MALDI-generated ions provides the advantage of utilizing the same instrument for a variety of applications, such as electrospray, AP-CI, and AP-MALDI.

The dependence of analyte chemical nature on an analyte/matrix cluster formation under AP-MALDI ionization conditions has been studied in some detail. Cluster formation has been shown to increase drastically with an increase in the number of arginine residues present in a given peptide. The ion trap instrument provides the additional benefit of effectively declustering AP-MALDI-generated ions by use of the tunable electrical field ion heating by increasing the potential between the skimmer and the first octapole to induce octapole ion source CID (Laiko et al. 2000).

2.2.6 SURFACE-ENHANCED LASER DESORPTION IONIZATION

The surface-enhanced laser desorption ionization (SELDI) technology was developed in 1993 (Hutchens and Yip 1993) and commercialized by Ciphergen Biosystems Inc. as the ProteinChip system which is now produced and commercially available from Bio-Rad Laboratories (Hercules, CA). SELDI may be considered as a variant of a MALDI technique. The difference is in the sample preparation step, where the matrix, in this case an energy-absorbing molecule (EAM), is mixed with the protein sample (Tang et al. 2003). In MALDI, a protein or peptide sample is mixed with the matrix molecule in solution. Small amounts of the mixture are "spotted" on a surface and allowed to dry. The peptide sample and matrix co-crystallize as the solvent evaporates. In SELDI, the protein mixture is spotted on a surface modified with some chemical functionality. Some proteins in the sample bind to the surface, while the others are removed by washing. After washing the spotted sample, a matrix (EAM) is applied to the surface and allowed to crystallize with the sample proteins/peptides. This technique utilizes stainless steel- or aluminum-based supports, or chips.

Sample interaction with a SELDI surface is used as a chromatographic step and consequently only a subset of proteins that bind to the given surface is retained, thus facilitating the analysis. Commonly used surfaces include weak cation exchanger, hydrophobic surface, immobilized metal affinity materials (metal-binding surface), and strong anion exchanger. Surfaces can also be functionalized with antibodies, other proteins, or DNA. Samples spotted on a SELDI surface are usually analyzed using TOF-MS. As mixtures of proteins will be analyzed within different samples, a unique sample fingerprint or signature will result for each sample tested. Consequently, patterns of masses rather than actual protein identifications are produced by SELDI analysis. Such mass patterns for biomolecules could be used to differentiate sample origin (Wright et al. 1999; Li et al. 2002) or to differentiate microorganisms.

2.2.7   DESI MS

DESI was developed by R. G. Cooks and coworkers from the Purdue University. This AP ionization technique combines features of ESI and desorption ionization methods by directing a stream of electrosprayed droplets of a suitable solvent onto a surface to be analyzed. Under these conditions, analytes on the surface undergo ionization and are directly transferred into a MS for analysis (Figure 2.5). Other

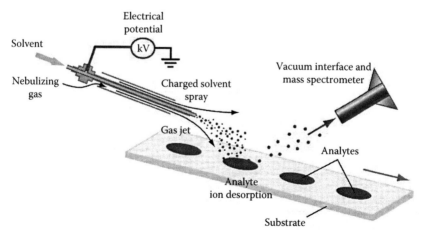

FIGURE 2.5   Schematic representation of DESI of analytes and direct transfer of analytes from the surface direct transfer into MS, for analysis. (From Cooks, R.G. et al., *Science*, 311(5767), 1566, 2006.)

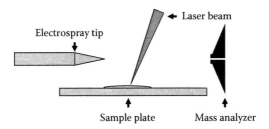

FIGURE 2.6   Schematic representation of ELDI technique.

methods in this group of ionization techniques include electrospray-assisted laser desorption ionization (ELDI), direct analysis in real time (DART), and the AP solids analysis probe (Cooks et al. 2006).

2.2.8   ELDI

Jentaie Shiea and colleagues at National Sun Yat-Sen University (Taiwan), the University of Portsmouth (UK), and Montana State University have developed an ambient MS ionization method called ELDI (Figure 2.6), the new technique that ionizes proteins and other molecules in a small, well-defined area on a solid surface for MS analysis without sample preparation. Because MALDI requires a matrix, which can interfere or complicate the analytical process, Shiea and colleagues developed a laser desorption method that does not use a matrix. A laser beam is shone on a surface, and molecules are desorbed. Many neutral species and few ions are produced with laser desorption, so ESI was used as a post-ionization method that uses fusion with charged solvent droplets generated by ESI.

The researchers tested ELDI on a sample of bovine cytochrome *c*. When laser desorption or ESI alone was applied to the sample, either no signal or only background ions were observed in the mass spectrum. When the two processes were combined to perform ELDI, however, many multiple-charged cytochrome *c* ions were obtained. This new, soft ionization method combines features of both ESI and MALDI. For instance, it works under ambient conditions and does not require any matrices; however, in comparison to DESI, a higher power used for desorption, therefore a broader range of molecules, can be analyzed and ESI-like ions are generated.

2.3   MASS ANALYZERS AND MS/MS METHODS USED FOR MICROBIAL DETECTION AND IDENTIFICATION

2.3.1   MASS ANALYZERS

Ionization of molecules produces molecular and fragments ions, which are formed in the source region of a MS and moved into a mass analyzer by an electric field. The mass analyzer separates these ions according to *m/z* values and detected ions are used to generate a mass spectrum providing molecular mass and structural information (Figure 2.7). The selection of a mass analyzer depends upon the resolution, mass

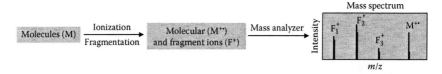

FIGURE 2.7   Measurement of m/z ratio of molecular and fragment ions based on detection of mass-analyzed ions by MS.

TABLE 2.2
Common Pulsed and Continuous Mass Analyzers Used in Biological MS

Pulsed Mass Analyzers	Continuous Mass Analyzer
Time of flight	Quadrupole
Quadrupole ion trap	Magnetic sector
Linear ion trap	
Orbitrap	
Ion cyclotron resonance	

---

range, scan rate, and detection limits required for an application. Each analyzer has very different operating characteristics and the selection of an instrument involves important trade-offs.

Analyzers are typically described as either continuous or pulsed. Continuous analyzers include quadrupole filters and magnetic sectors. These analyzers are similar to a filter or monochromator used for optical spectroscopy. They transmit only ions with single, selected m/z values to a detector and the mass spectrum is obtained by scanning the mass range, so different m/z ratio ions are detected. While a certain m/z is selected, any ions at other m/z ratios are lost, reducing the S/N for continuous analyzers. Single ion monitoring (SIM) enhances the S/N by setting the MS at the m/z for an ion of interest. Since the instrument is not scanned the S/N improves, but any information about other ions is lost.

Pulsed analyzers are the other major class of mass analyzers that include TOF and ion trap-MS, that is, quadrupole and linear ion traps (LIT) as well as an ion cyclotron resonance (ICR) and orbitrap-MS (Table 2.2).

2.3.1.1   TOF Mass Analyzers

TOF mass analyzers have in principle an unlimited mass range and high sensitivity because all ions can be recorded without scanning. A TOF mass analyzer consists of an ion source, a field-free drift chamber held under high vacuum, and a detector (Figure 2.8a). Although ions generated in the source have different masses, they are accelerated to the same KE and as a result they drift with different velocities and reach a detector at different times. In short, ions arrive at the detector with the KE obtained from the potential energy of the electric field in the source. Since KE

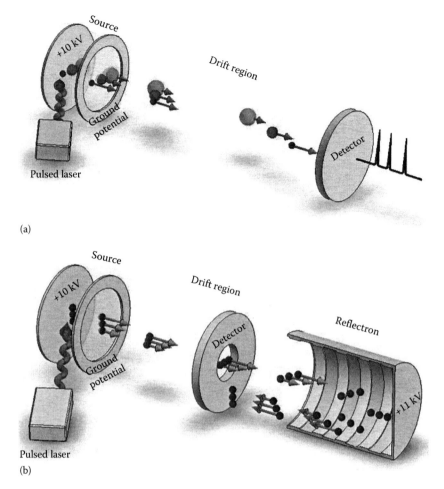

(a)

(b)

FIGURE 2.8 Schematic of (a) simple linear TOF-MS illustrating how ions separate in time owing to their differences in mass and (b) a reflectron TOF-MS illustrating energy-focusing effect on a packet of single-mass ions with small differences in KE (ions are on the vertical plane). (From Ecelberger, S.A. et al., *Johns Hopkins APL Tech. Digest*, 25(1), 14, 2004.)

of ions with mass m is equivalent to potential energy of the electric field and is equal to $mv^2/2$, ions travel with velocity v that is equal to $(2KE/m)^{1/2}$. Ions with the same KE and different masses traverse the analyzer within a time that depends on their m/z values. A detector positioned at the end of the analyzer drift tube measures the arrival time of ions that allows calculating their masses. Ions with lower m/z will have greater velocity than ions of greater m/z (Skoog et al. 1992; Guilhaus 1995). Although the resolving power of simple TOF instruments is low, they provide the opportunity to measure masses of biomolecular complexes that include even whole viruses.

Two approaches have been developed to increase mass resolution and accuracy of TOF analyzers. The first one is referred to as delayed extraction, where a time delay is introduced between ionization and the extraction potential. However, optimal focusing can only be achieved for a narrow *m/z* range. The second approach is the introduction of a series of evenly spaced electrodes, a reflectron, at the end of the linear flight tube (Mamyrin 2001).

2.3.1.1.1   Delayed Ion Extraction Linear TOF-MS

To improve mass resolution of a TOF analyzer, a technique named the delayed ion extraction (Vestal et al. 1995) was developed, which can compensate for the initial ion velocity distribution in the source. In a delayed ion extraction mode of operation, MALDI-generated ions are produced in a transient field-free region in the ion source and the plume is allowed to diffuse for a fraction of a microsecond. Creating an electric field in the ion source with a fast high-voltage pulse then accelerates the ions. The time delay and extraction potential can be set so that all ions with the same *m/z* are focused at a certain plane in space. In other words, this pulse voltage and time delay corrects the initial velocity component of MALDI-generated ions that results in a simultaneous arrival of ions with identical *m/z* at the detector. This results in narrower ion arrival time distributions and provides better mass resolution when compared to the standard ion extraction mode of operation.

2.3.1.1.2   Reflection TOF-MS

The most popular method developed for improving mass resolution in MALDI-TOF-MS uses an attached single-stage or a dual-stage reflectron (RETOF-MS). The ion reflectron is used to compensate for differences in TOF of the same *m/z* ions due to slightly different kinetic energies. A reflectron TOF-MS analyzer is shown schematically in Figure 2.8b, which illustrates the energy-focusing effect on a packet of single-mass ions with small differences in KE. In this simplest, single-stage type of a reflectron, a homogeneous reflecting electrostatic field is created between two parallel flat grids. The first grid constitutes the entrance of the ion mirror, and the second, the end. In order to improve the homogeneity of the field in the reflectron, a number of equally spaced ring electrodes are usually placed between the end electrodes connected by a chain of equal value resistors. The second generation of a reflectron TOF analyzer is a double-stage reflectron with two separate homogeneous field regions of different potential gradients. Various studies have shown that by choosing suitable mirror voltages and dimensions to match the lengths of the field-free region, a large enhancement in resolution can be achieved. In particular, a spectacular resolution gain can be realized for ion beams with broad KE distributions (Karataev et al. 1972; Mamyrin et al. 1973).

The quadratic field reflectron (Z^2) is an ion mirror reversing the trajectory of incoming fragment ions and their intact precursor ions that is of paramount importance for studying the fragmentation process. In this analyzer, smaller fragment ions, which possess relatively low KE, do not penetrate the reflectron field to the same extent as the larger precursor ions and are consequently reflected onto a new trajectory, arriving at the detector prior to the larger ions. Since the fragment and precursor ions are detected at different times, high-resolution separation of the different ion groups is achieved. The quadratic field reflectron keeps its energy-focusing properties independently of the KE (i.e., penetration depth) of the ions entering.

Therefore, fragment ions that originate from a selected parent ion and cover a broad range of kinetic energy can be focused in a single analysis, while a traditional linear reflectron can only focus ions with a narrow range of kinetic energies. Therefore, a complete post-source decay spectrum can be obtained in 1–2 min on an instrument with a quadratic field reflectron in comparison with ~30 min required when a linear reflectron TOF instrument is applied for acquiring a post-source decay spectrum.

2.3.1.2 Magnetic Sector Analyzers

Magnetic field was used in the earliest mass analyzers to separate ions. In this approach, ions are accelerated and pass through a magnetic field applied in a direction perpendicular to their motion. Under such circumstances, ion velocity remains constant and travels in a circular path. A curvature of the path depends upon the m/z of the ion in the beam and is influenced by differences in initial velocity of ions. In the magnetic sector analyzer, the trajectory of the ions is usually controlled by a variable magnetic field allowing ions of interest to be sequentially collected onto a single detector. However, the resolution is limited because not all ions leaving the ion source have exactly the same energy and velocity.

To improve the resolving power of an analyzer, double-sector instruments have been developed to combine the magnetic (B) sector and electrostatic (E) analyzers. The electric sector acts as a KE filter and allows only ions of uniform KE to reach the detector. Overall, these instruments are designed to provide resolving power on the order of 10^5 and allow separation of various ionic species with the accuracy of a few parts per million (ppm). The main limitations of magnetic sector are its unsuitability for pulsed ionization methods; moreover, they are usually expensive in comparison to other mass analyzers and provide limited precursor selectivity with unit product-ion resolution (Trainor and Derrick 1992; Chen 1997; Wang et al. 1999a,b). Nevertheless, these double-focusing magnetic sector mass analyzers are the classical models against which the performance of other analyzers is compared.

2.3.1.3 Quadrupole Mass Analyzer

The quadrupole MS is the most common and relatively inexpensive mass analyzer. This is due to its compact size, rapid scan rate, efficient ion transmission, and modest vacuum requirements, which are ideal for small instruments. The quadrupole mass analyzer is considered as a "mass filter" because through employing concurrent DC and radio-frequency (RF) potentials on the quadrupole rods, only ions with a selected m/z ratio are allowed to pass to the detector. All other ions do not have a stable trajectory through the quadrupole mass analyzer and will collide with the quadrupole rods, never reaching the detector. The resolution of most quadrupole mass filter instruments is limited to unit m/z and has a mass range of m/z 1000. Many benchtop instruments have a mass range of m/z 500 but research instruments are available with mass range up to m/z 4000. In the MS, an electric field accelerates ions out of the source region and into the quadrupole analyzer.

The quadrupole analyzer consists of four rods or electrodes arranged across each other with a direct current voltage and superimposed RF potential (Figure 2.9).

Theoretical treatment of quadrupole operations was reviewed by Todd (2005). In short, the RF and DC voltages applied to the quadrupole rods determine the ion m/z

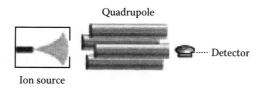

FIGURE 2.9   Schematic representation of a quadrupole mass filter.

value transmitted by the quadrupole. The four rods are connected in pairs and the RF potential is applied between these two pairs of electrodes. During the first part of the RF cycle, the top and bottom rods are at a positive potential and the left and right rods are at a negative potential. This squeezes positive ions into the horizontal plane. During the second part of the RF cycle, the polarity of the rods is reversed. This changes the electric field and focuses the ions in the vertical plane. The quadrupole field continues to alternate as the ions travel through the mass analyzer. This causes the ions to undergo a complex set of motions that produces a three-dimensional (3D) wave. Selected ions are transmitted through the quadrupole field because the amplitude of this 3D wave depends upon the m/z value of the ion, the potentials applied, and the RF. The RF and DC fields are scanned to collect a complete mass spectrum (Dawson 1976).

Quadrupole analyzers are broadly used in benchtop GC-MS and LC-MS platforms and as components of hybrid instruments like triple quadrupole (TQ), sector-quadrupole, or quadrupole-TOF (Q-TOF) analyzers.

2.3.1.4   Ion Traps

There are two principal ion-trapping mass analyzers: quadrupole ion traps (QIT) (dynamic traps) and ICR-MS (static traps). Both operate by storing ions in the trap and manipulating the ions by using DC and RF electric fields in a series of carefully timed events.

2.3.1.4.1   QIT

A QIT is a 3D, dynamic ion storage device (Figure 2.10). It consists of three electrodes, namely, two end caps and a ring electrode. Current analytical use of ion traps relies on storing ions and then ejecting them in a mass-selective manner. This approach was introduced commercially by Finnigan Mat (now Thermo-Fisher, San Jose, CA) in their "Ion Trap Detector" MS in the middle 1980s. In accordance with this approach, QIT is operated by introducing ions through one of end caps

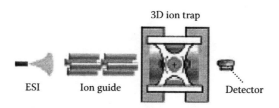

FIGURE 2.10   Schematic representation of a 3D QIT equipped with the ESI ion source.

(or generating them directly in the trap by EI or CI methods) and applying a sinusoidal potential at RF to the ring electrode, while keeping the end caps grounded. Ions with a broad range of m/z values can be held in stable orbits by a superposition of appropriate DC potential on top of the RF-drive potential. However, by increasing the amplitude of the RF potential, the motion of some ions becomes unstable and they develop unbound trajectories along the axis of symmetry. Such ions exit the ion trap through the holes in the end cup and strike a detector (e.g., electron multiplier) in the order of their m/z values. Usually, a few m/z segments and microscans are recorded separately and displayed as a final mass spectrum.

Further improvements were achieved through the technique of axial modulation that is by applying a supplementary oscillating field of a few volts at a frequency of about half of the RF-drive potential between the end-cap electrodes. This technique extends the m/z range of the trap, while the application of the supplementary oscillating field is also the basis of studying the collision-induced dissociation (CID) of ions in the trap, known as tandem in time and MS^n capabilities. Nevertheless, space-charge effects initially limited the inherent dynamic range of the ion trap. Therefore, the idea of automatic gain control was introduced that incorporates two ionization stages into a scan function. In short, an additional pre-scan is performed to determine the total ion signal that is next used to calculate the optimum ionization time for the second, analytical ionization event.

Overall ion traps have experienced a broad range of applications, due to their high sensitivity and selectivity; ability to perform multistage MS experiments (MS^n); and compact sizes. Ion traps have been successfully interfaced with many ionization sources, i.e., AP ionization, ESI, and laser desorption (Barinaga et al. 1994; Doroshenko and Cotter 1994, 1996). Moreover, it is the analyzer of choice in many hybrid MS configurations, i.e., QIT-TOF, magnetic sector-QIT, etc. (Fountain et al. 1994; Zerega et al. 1994; Qian and Lubman 1995; Jonscher and Yates 1996; Todd 2005).

2.3.1.4.2 LIT

Linear, or two-dimensional (2D), ion traps (LITs) are relatives of QIT mass filters that resemble quadrupoles, except that additional DC potentials allow for trapping of ions along the long axis. LIT is a square array of four hyperbolic rods, where rods opposite to each other are connected electrically. Thus the four rods can be considered to be pairs of two rods. Scanning is performed through a ramping protocol that either ejects ions radially (e.g., LTQ, Thermo-Fisher, Figure 2.11) or axially

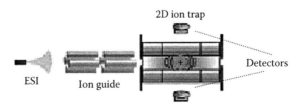

FIGURE 2.11 Schematic representation of linear (2D) ion trap mass analyzer (LTQ) manufactured by Thermo-Fisher (ESI, Electrospray Ionization Ion Source).

FIGURE 2.12 Schematic representation of a LIT (Q-trap) mass analyzer manufactured by ABI/MDX-Sciex.

(e.g., Q-trap, ABI/MDX-Sciex, Figure 2.12). A major advantage of the 2D traps is that they possess greater trapping volumes and can thus analyze more ions per cycle, with concomitant improvements in sensitivity and dynamic range.

The 2D LIT in the LTQ is comprised of four parallel hyperbolic shaped rods, segmented into three sections. Ions are trapped radially in a RF electric field and axially in a static electric field using DC voltages. Application of appropriate voltages to all three segments generates a homogeneous field throughout the trapping region. Mass analysis using the Finnigan LTQ involves ejecting the trapped ions in the radial direction through the two parallel slots in the center section of the LIT. Two highly efficient detectors are placed on either side of the trap to maximize ion detection sensitivity.

The trap works by trapping ions through the application of three different DC axial trapping voltages to the front, center, and back sections of the analyzer aimed to hold the potential of the center section below the potential of the front and back sections. These ions are also prevented from escaping along the z-axis by putting a fixed positive voltage on the pin-like electrodes at either end of the trap. In addition, AC voltages of the same amplitude and sign are applied to the rods of each pair, while AC voltages applied to different pairs are equal in amplitude but opposite in sign. The AC voltages oscillate with a constant RF (e.g., 1.2 MHz) and of variable amplitude (0–10,000 V). This time-varying field drives ion motion in the radial direction and may provide stable trajectories for ions characterized by a broad range of m/z values. It takes place when the amplitude of the RF voltage is low; however, when this voltage increases, ions with the increasing m/z values become successively unstable and are ejected from the mass analyzer and used to reconstruct a mass spectrum.

By the application of ion isolation waveform voltages, resonance excitation RF voltages, and resonance ejection RF voltages to exit rods in a timely manner, only selected ions can be isolated, fragmented by collisions with the helium damping gas, and mass analyzed. Altogether, these processes allow for many applications like tandem-in-time MS experiments (MS/MS), selected ion monitoring (SIM), selected reaction monitoring (SRM), or consecutive reaction monitoring by using MS^n scan mode of operations.

2.3.1.5 Fourier Transform ICR Mass Analyzer

Fourier transform ICR (FT-ICR) MSs measure mass indirectly by oscillating ions in a strong magnetic field. While the ions are orbiting, an RF signal is used to excite them. Because these ions will oscillate as a function of their m/z, measuring the frequency of these oscillations allows inference of m/z by using a FT. These instruments provide the highest mass resolution of all MS in combination with highest mass accuracy. However, they are limited to expert-only laboratories due to the need for a large superconducting magnet and concomitant requirement for liquid helium and nitrogen.

2.3.1.6 Orbitrap Mass Analyzer

The orbitrap also uses an FT-based strategy to measure m/z ratios of ions. However, the trapping is performed electrostatically (as opposed to magnetically) and the frequency oscillations are measured along the long axis of the trapping cell. Moreover, orbitrap is an effective MS with mass resolution surpassed only by FT-ICR and achieves it by using a much simpler, compact design. It operates by radially trapping ions about a central spindle electrode.

In the absence of any magnetic or RF fields, ion stability in this trap is achieved only due to the balance between centrifugal and electrical forces. The former requires ions to orbit around an axial electrode, hence the name orbitrap for the technique. At the same time, ions also perform harmonic ion oscillations along the electrode. These oscillations are detected using image current detection and transformed into mass spectra using fast FTs as in FT-ICR.

The potential advantages of the orbitrap include (a) a high mass resolving power (up to 200,000), (b) increased space charge capacity at higher masses due to independence of trapping potential and larger trapping volume in contrast to FT-ICR and quadrupole traps, (c) high mass accuracy (1–2 ppm), and (d) high dynamic range (around 5000) (Makarov et al. 2006a,b).

2.3.2 METHODS OF ION EXCITATION AND ANALYSIS OF PRODUCT IONS WITH MULTIPLE MASS ANALYZERS

The utilization of MS analyzers in tandem is another approach to obtaining mass spectrum of a precursor ion that undergoes dissociation to products. In order for this process to occur, a significant amount of the precursor ions should undergo some form of dissociation between two stages of mass selection. There are numerous modes of dissociation including, but limited to, metastable dissociation, CID, and electron-induced dissociation.

2.3.2.1 Ion Excitation

One of the primary technological advances that has made possible in-depth analysis of ions, like sequencing of peptide ions, is developments in the ability to excite, fragment, and analyze product ions. In the simplest form, that is, during tandem MS analysis (MS/MS), a precursor ion of interest is isolated, activated, fragmented, and the resulting product ions analyzed. Fragmentation of precursor ions in MS/MS is accomplished either through application of "slow heating" methods or of more recently developed gas-phase ion–ion reactions or ion–electron reactions. "Slow heating" ion activation methods such as low-energy CID and infrared multiphoton dissociation (IRMPD) involve slow addition of energy to the precursor ions such that the internal energy of the ions exceeds the energy needed for fragmentation (Figure 2.13a) (Shukla and Futrell 2000; Sleno and Volmer 2004). The most commonly used slow heating fragmentation method is low-energy CID (also known as collisionally activated or aided dissociation, or CAD). This method is employed in hybrid instruments like TQ, Q-TOF, QIT, LIT and in the ICR cell of an FT-ICR.

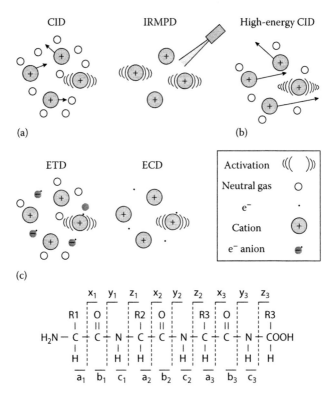

FIGURE 2.13   Ion activation methods for peptide or protein MS/MS. (a) Slow heating methods CID: most commonly used in ion traps, CID is a process whereby precursor ions are reacted with a neutral and inert gas, such as argon. Each collision with neutral gas increases the overall vibrational energy of the molecule until enough internal energy is accumulated within the molecule to overcome the bond energy, thus causing dissociation. IRMPD: most commonly used in ICR cells, IRMPD uses a laser pulse to induce vibrational energy in the molecule. Absorption of photons causes vibrational energy in the molecule to increase incrementally until bond energy overcomes and dissociation occurs in much the same manner as CID. (b) High-energy CID: commonly used in TOF–TOF instruments, high-energy CID works in the same manner as CID with the primary difference being that peptide precursor ions enter the collision chamber at high energies (typically 1–2 keV), so that only a single collision with gas is needed to cause fragmentation. High-energy CID results in peptide backbone fragmentation yielding b and y ions, and generates immonium ion fragments and internal d and w fragment ions. (c) Ion–ion and ion–electron reactions ETD: used exclusively in ion traps, ETD is an ion–ion reaction between single-charged anions, which also serve as a source of electrons, and multiple-charged cations. The anions are allowed to interact with the multiple-charged peptides and result in proton transfer without dissociation as well as electron transfer with or without dissociation. Proton transfer results in charge reduction while dissociation leads to c and z fragment ions. ECD: used in FT-ICR instruments, ECD relies on interaction of precursor ions with low-energy electrons introduced into the ICR cell. The capture of low-energy electrons by the multiple-charged protein or peptide precursor leads to neutralization and dissociation of the ion. Backbone cleavage occurs rapidly at or near the site of electron capture, resulting in c and z fragment ions as in ETD. (From Khalsa-Moyers, G. and McDonald, W. H. 2006. Developments in mass spectrometry for the analysis of complex protein mixtures. *Briefings in Functional Genomics and Proteomics* 5:98–111.)

2.3.2.1.1 CID

In a CID experiment, precursor ions are isolated and subjected to collision with a neutral, inert gas. The energy imparted to the precursor ion upon collision is converted from KE to internal energy. Multiple collisions result in an increase in internal energy of the precursor ion that causes its fragmentation; for example, in case of peptide or protein ions, their backbone fragments, mainly at amide bonds, resulting in the formation of highly specific series of ions (e.g., b- and y-type fragment ions). Within ion traps and ICR cells, the collision is considered to be tandem in time because the initial isolation and subsequent activation and fragmentation steps occur in the same location. Tandem-in-space experiments are performed in TQ, Q-TOF, or Q-FTMS instruments. Precursor ions enter the first quadrupole, and selected ions are sent to the second quadrupole, where they are focused by the application of an RF-only voltage. The collision cell is filled with neutral gas, and ion activation results from multiple collisions of precursor ions with the gas. Resulting fragments are then measured in the downstream detector (e.g., TOF in the Q-TOF instrument). Another instrument that performs a tandem-in-space MS/MS is the TOF–TOF. However, because the ions have higher energies during the fragmentation process, it produces somewhat different MS/MS fragmentation patterns and is considered a high-energy CID process (Figure 2.13b).

2.3.2.1.2 IRMPD

IRMPD is an alternative slow heating dissociation method involving nonresonant ion activation and subsequent dissociation via photon absorption. Historically used in analysis of small molecules, IRMPD is now becoming more popular in analysis of larger peptides and proteins. In an IRMPD experiment, ions are typically activated with a low-power CO_2 laser of wavelength 10.6 mm. Since it is also a slow heating method, IRMPD gives nearly identical fragment ions to those seen with dissociation by CID (Zhang et al. 2005) (Figure 2.13a). Complementary to the slow heating ion activation methods in MS/MS are the ion–electron and ion–ion reactions (Figure 2.13c).

2.3.2.1.3 Electron Capture Dissociation and Electron Transfer Dissociation (ETD)

Electron capture dissociation (ECD) is based on ion–electron reactions in which the capture of electrons by a gaseous positive ion leads to fragmentation and neutralization of the positive ion (Shukla and Futrell 2000). When applied to multiple-charged peptide or protein cations, ECD results in extensive cleavage of the backbone at N-Cα bonds to yield c and z ions (Figure 2.13) (Zubarev et al. 1998). ECD also preferentially cleaves disulfide bonds but leaves other posttranslational modifications (PTMs) such as sulfonation and gamma-carboxylation intact (Zubarev 2004). Ion residence time in ICR cells is typically greater than that in ion traps, making FT-ICR better suited to ECD. Additionally, ECD efficiency is greatest for electron energies less than 1 eV; however, such energies are difficult to achieve in an ion trap instrument due to the strong RF potentials needed to trap ions. Because ECD provides complementary information to slow heating activation, ECD is an available option on all commercial FT-ICR instruments and has been applied in the structural characterization of numerous PTMs (Bakhtiar and Guan 2005). Since ECD is not amenable to ion trap instruments, an analogous ion–ion technique was developed by Hunt and colleagues (Syka et al. 2004a,b) and is an extension of earlier electron transfer reaction studies in nucleic acids done by McLuckey and colleagues (Wu and McLuckey 2003). Electron transfer dissociation (ETD) is an ion–ion reaction between single-charged anions and multiple-charged cations, like peptide ions. In an ETD experiment, reagent anions are created by a CI source and serve as a source of electrons. The reagent anions are introduced into the trap and are allowed to interact with the multiple-charged peptides. This results in proton transfer without dissociation as well as electron transfer with or without dissociation. Proton transfer results in charge reduction, while dissociation leads to c- and z-fragment ions and gives the extensive peptide backbone fragmentation characteristic of ECD.

2.3.2.2 Geometry and Operation of Hybrid Mass Analyzers

Advanced MS methods increasingly involve the use of MS/MS (see Figure 2.14). In these experiments, both parent and product ions must be independently mass analyzed and hence the experiment requires two stages of mass analysis. These stages of analysis can be performed using any two of the analyzers, e.g., two quadrupoles (QQ), or a Q-TOF ("tandem in space"). However, some instruments allow the parent and product ions to be analyzed using a single device, which is used sequentially ("tandem in time").

MS/MS experiments provide a form of separation, which makes them particularly useful for complex mixture analysis. They are also useful, however, in increasing the specificity of identification because the technique provides additional characteristics (fragment ions derived from a particular parent ion) by which the compound can be identified.

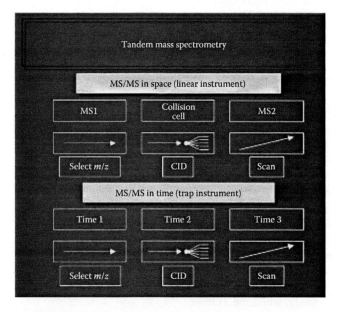

FIGURE 2.14 MS/MS illustrating tandem-in-time and tandem-in-space concepts.

2.3.2.2.1   TQ

TQ-MS allows for MS/MS experiments and consists of three sections (Q1, Q2, and Q3). Q1 acts as a mass filter allowing only precursor ions of a certain mass to move further into the Q2, which operates in an RF mode only and functions as a collision cell with helium atoms for fragmentation of the ions. Q3 acts as a second mass separating quadrupole, allowing the fragment ions to be separated and resolved before they are measured by the detector (Yost and Enke 1979, 1983).

Scanning the first quadrupole and holding the other two quadrupoles in an ion transmission mode gives a precursor-ion spectrum. Then as ions are selected they could be collided in the second quadrupole while the other two are at constant m/z; once product ions are generated, they are transmitted to the third quadrupole for obtaining the mass spectrum of the selected precursor ion. TQ instruments are characterized by a mass accuracy of 0.01%, low detection limit (~2 pmol), provide true precursor and neutral loss analysis capabilities, and can be easily interfaced to soft ionization ion sources.

2.3.2.2.2   Quadrupole/LIT/Quadrupole (Q-LIT-Q)

The combination of TQ-MS with LIT technology in the form of an instrument of configuration shown in Figure 2.14 uses the classical TQ scan functions that allow for SRM, product ion, neutral loss, and precursor ion scanning in addition to sensitive ion trap experiments. Moreover, quantitative and qualitative analysis can be performed using the same instrument. Furthermore, for peptide analysis, the enhanced multiple-charged scan allows an increase in selectivity, while the time-delayed fragmentation scan provides additional structural information. Various methods of operating this hybrid analyzer are described for the commercial instrument (Q-trap, AB/MDX-Sciex) that is suitable for analysis of microbial constituents, including proteomics applications (Hager and Le Blanc 2003; Hopfgartner et al. 2003, 2004; King and Fernandez-Metzler 2006).

The commercial QTRAP® LC/MS/MS system is a high-performance hybrid TQ-LIT-MS that enables high-sensitivity MS/MS, and MS³ experiments as well as quantitative analysis to be performed within a single platform. In addition, the use of a quadrupole as a LIT significantly enhances ion trap performance by increasing ion capacity, improving injection, and trapping efficiencies. It covers a mass range from m/z 50 to 2800 with a scan speed of up to 4000 amu/s, and the mass resolution better than 3000 as well as good mass stability (0.1 amu over 9 h at m/z 2010) has been observed with this instrument at normal operating temperature.

2.3.2.2.3   Qq-TOF

The Qq-TOF tandem MS can be thought of as a TQ with the last quadrupole section replaced by a TOF analyzer (Figure 2.15). This analyzer consists of two quadrupoles, followed by a reflecting TOF mass analyzer with orthogonal injection of ions. The second quadrupole is introduced to provide collisional dissociation of ions. In some commercial instruments, the quadrupoles may be replaced by hexapoles; however, the basic operating principles are the same (Morris et al. 1996).

For a single MS (or TOF-MS) measurement, the mass filter Q1 is operated in the RF-only mode, while the TOF analyzer is used to record spectra. The resulting spectra benefit from the high resolution and mass accuracy of the TOF instruments,

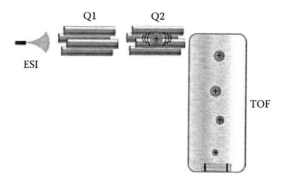

FIGURE 2.15   Schematic diagram of a tandem QqTOF-MS.

and also from their ability to record all ions in parallel, without scanning. For an MS/MS scan, Q1 is operated in the mass filter mode to transmit only the parent ion of interest. The ion is then accelerated to an energy of between 20 and 200 eV before it enters the collision cell (Q2), where it undergoes CID after the first few collisions with neutral gas molecules (Ar or N_2). The resulting fragment ions (in addition to the remaining parent ions) are collisionally cooled and focused as described earlier. This step is even more important in Qq-TOF instruments than it is in TQ because the TOF analyzer is much more sensitive to the "quality" of the incoming ion beam than is Q3 in a TQ instrument.

These analyzers provide attomole-level sensitivities and very good mass accuracy (better than 2 ppm) that increases specificity and confidence in results. Moreover, Qq-TOF analyzers are characterized by other attractive features like broad dynamic range (3–4 orders of magnitude) and fast MS and MS/MS spectra acquisition time.

2.3.2.2.4   LIT-FTICR Analyzer

The LIT-FTICR hybrid system usually consists of an AP-ESI source, an RF multipole-based ion accumulation and transportation system, and a permanent magnet FTICR mass analyzer. A magnet generates an axial, solenoid-type magnetic field with homogeneity below 50 ppm over a central 1 cm^3 region. One of the RF quadrupoles is segmented and works as a LIT for ion accumulation, selection, and collision-induced fragmentation (Martin et al. 2000; Syka et al. 2004a,b). These instruments provide very good mass accuracy and the resolution of greater than 500,000 that guarantee unprecedented LC-MS performance and sub-femtomolar sensitivities. Unfortunately, the use of these instruments is limited to expert-only laboratories due to the need for a large superconducting magnet and concomitant requirement for liquid helium and nitrogen.

2.3.2.2.5   LIT-Orbitrap Hybrid Mass Analyzer

The hybrid LIT-orbitrap mass analyzer is a FT-type instrument that combines a LIT and an orbitrap analyzer that is suitable for a wide variety of different experiments. In this hybrid instrument, ions are collected in the LIT and transferred by axial ejection to the C-shaped storage trap, which is used to store collisionally cool ions before

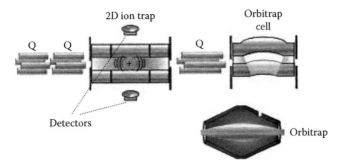

FIGURE 2.16   Schematic representation of LIT-orbitrap hybrid mass analyzer.

injecting them into the orbitrap. The ions transferred from the C-trap are captured and analyzed in the orbitrap as previously described (Figure 2.16). The commercial hybrid system, LTQ-Orbitrap XL (Thermo-Fisher), features a new collision cell to provide additional flexibility to any MS^2 experiment. This system is considered as a very important tool for proteomics research (Hu et al. 2005). For example, Macek et al. (2006) used an orbitrap instrument to accurately measure the mass of intact standard proteins; the LIT was used to collisionally dissociate various charge states of the protein, followed by injection of these ions into an orbitrap to determine the masses and charge states of the fragment ions.

The analysis of complex protein mixtures is one of the most challenging tasks in the area of proteomics. The key requirements of MS-based methods are a wide dynamic range, outstanding mass accuracy, fast cycle times for MS/MS experiments,

TABLE 2.3

Comparison of Instrumental Capabilities for Various Hybrid Mass Analyzers

	TQ (QqQ)	Q-LIT-Q	Qq-TOF	LIT-FTICR	LIT-Orbitrap
Resolution	0.01%	>3000	2 ppm	2 ppm	<2 ppm
Sensitivity	2 pmol	10–100 fmol	Attomole level	Sub-femtomole	Sub-femtomole
Ionization source; MS^n	ESI/MS^2 API/AP-CI	ESI/API/ AP-CI; MS^3	$ESI/MALDI/MS^2$	ESI/API; MS^n	ESI/API; MS^n
Mass range	m/z 50–2500	m/z 50–3000	m/z 500–100 K	m/z 50–5000	m/z 50–3000
Mass accuracy	Medium	Medium	Good	Excellent	Very good
Posttranslational modifications	None	Excellent	Low	Low	Excellent
Fieldable potential	Need modification	Not available	Some prototypes are available	Not available	Not available
Price range	$295–380 K	>$350 K	$400–480 K	>$480 K	$400–600 K
Capillary LC compatibility	Yes	Yes	Yes	Yes	Yes

and high sensitivity. The LTQ-Orbitrap XL meets all of these requirements and provides the most confident protein identification.

Advantages of the orbitrap-LIT mass analyzer include very high mass accuracy, sensitivity, and dynamic range. However, most importantly this analyzer obviates the need for high field–strength magnets and their cryogenic requirements.

A comparison of instrumental capabilities for various hybrid mass analyzers is given in Table 2.3. Each has its favorable attributes and likewise limitations. These differences are suitable and optimized for different and various applications. It is important in the context of microbe detection and identification to consider the level of detection and identification when looking at these types of charts. An application to identify simply microbes is different than an application to identify microbes at strain level.

3 Matching Mass Spectral Profiles of Biomarkers

Michael F. Stanford

CONTENTS

3.1   INTRODUCTION

MS applications were derived from many sources. The perception held by many users was that this technology was big, heavy, required specialized handling, and most importantly required high vacuum to work. Also, MS methods belonged to the chemists and most early instruments were in chemistry departments and only recently have biologists been allowed access. This thinking was held and is still held by many outside the discipline. The result of this thinking was the development of many different approaches to solving these perceived issues. Many interesting instruments and approaches resulted and are presented here.

3.2   MALDI-MS

With the development of MALDI-MS, this technique has been widely employed in the analysis of bacterial constituents, including proteins (Krishnamurthy et al. 1996; Van Baar 2000). In particular, a substantial body of published reports focused

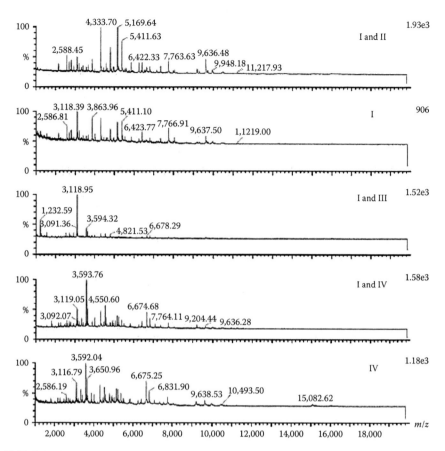

FIGURE 3.1   MALDI mass spectra of *B. anthracis* treated with different sample solvents. The matrix solution CHCA was dissolved in acetonitrile–methanol–water (1:1:1) with 0.1% formic acid and 0.01 M 18-crown-6. The other solvent treatments, including II, III, IV, and a combination of I and V, gave no signal (data not shown). I, 0.1% TFA; II, chloroform–methanol (1:1); III, 2-propanol–acetonitrile (1:1); IV, formic acid–2-propanol–water (1:2:3). (From Liu, H. et al., *Appl. Environ. Microbiol.*, 73(6), 1899, 2007.)

on the use of MALDI-TOF-MS (Figure 3.1) for near real-time discrimination between bacterial species (Fenselau and Demirev 2001; Lay 2001; Liu et al. 2007). Initial studies represented chemotaxonomic approaches focused on using patterns of masses deduced from MALDI spectra for bacterial strain identification. Such "mass fingerprints" are dominated by singly ionized proteins; therefore, they were usually matched against reference libraries containing experimentally determined protein masses (Bright et al. 2002) or theoretical protein masses calculated from genomic sequences of microorganisms (Demirev et al. 2001a,b). Although the latter approach represents an effort aimed at linking a genome and its corresponding proteome by MS, PTM and sample processing–related modifications might change the mass of genome-predicted polypeptides and therefore do not allow for the reliable identification of proteins. However, strain-specific protein mass profiles revealed by

MALDI-MS still provide an attractive way to generate phenotypic characteristics that are suitable for typing investigated strains. Because these approaches typically involve training of an expert system on relatively small sets of organisms, it is difficult to predict the general applicability of the identified biomarkers.

3.2.1 SAMPLE PREPARATION AND INSTRUMENTAL PARAMETERS FOR MALDI-MS

The processing of bacterial samples is very important (see Chapter 4) because it influences the quality of spectra and robustness of any data analysis algorithm. The following presents sample and instrumental parameters that were investigated to obtain high quality and reproducible mass spectra of bacterial constituents that are dominated by proteins.

Three general processing procedures are used to generate protein ions for subsequent characterization and identification of bacteria with MALDI-MS. The simplest of them uses a mixture of the bacterial sample with a matrix deposited onto a metal MALDI target (Amado et al. 1997; Wang et al. 1998; Gantt et al. 1999; Lee et al. 2002; Williams et al. 2003a,b; Hettick et al. 2004; Ruelle et al. 2004). The second method consists of suspending whole cells in a solvent capable of extracting proteins from the bacterial sample. A portion of the protein extract is mixed with matrix and analyzed by MALDI-MS. The third method uses lysis techniques to fragment the bacterial cell followed by the solvent extraction of cellular proteins. A protein extract is mixed with a matrix and the mixture is analyzed by MALDI-MS to produce a MALDI mass spectrum that represents percentages of protein species in the sample spot.

The matrix itself has been intensely studied with regard to its efficiency in transferring energy because chemical constituents in the matrix determine the extent of protein ionization and the dynamic mass range of the ions in the mass spectrum. A number of reports focus on the laser power, intensity, and laser spot size in the efficient desorption and ionization phenomena that fundamentally characterize the MALDI process (Cain et al. 1994; Williams et al. 2003a,b; Wunshel et al. 2005).

3.2.2 ANALYSIS OF MALDI-MS SPECTRA

Diverse data analysis methods are used to transform the MALDI mass spectral data into meaningful information. Initially, the simplest form of data analysis performed is a visual determination of the replicate mass spectra of the same organism and a comparison of spectra from different organisms. This necessarily requires that either some masses are unique to a given bacterium relative to a set of different bacterial mass spectra, or the intensity distribution is markedly different for the same set of masses between different bacterial mass spectra. For instance, Cain et al. (1994) showed the presence of similar masses but at significantly different intensities for different species of *Pseudomonas*. Holland et al. (1996) showed that the spectra of the *putida*, *aeruginosa*, and *mendocina* species of *Pseudomonas* have similar as well as different masses. Krishnamurthy et al. (1996) showed that protein extracts of *B. anthracis* Sterne, *Bacillus thuringiensis*, and *Bacillus cereus* displayed similar mass spectral masses and intensities. Haag et al. (1998) presented distinctly different

MALDI mass spectra for four different *Haemophilus* species. Differentiation of each culture was feasible by visual analysis of the mass spectral fingerprints.

Arnold and Reilly (1998) took the MALDI mass spectral analysis of bacteria one step further and essentially automated the manual, visual fingerprint approach. Mass spectra were evaluated in pairs by the standard cross-correlation method. However, the mass spectrum of a bacterium was divided into mass intervals and each interval was cross-correlated with that of a different bacterial spectrum. A spectrum from *m/z* 3,500 to 10,000 was divided into 13 intervals consisting of 500 Da in each interval. The product of the 13 cross-correlation values between a pair of spectra defined the final composite correlation index. Visually, similar spectra were differentiated using the cross-correlation technique for strains of *E. coli*. A fundamental tenet of microbiological taxonomy states that many strains of a bacterial species should be investigated for a satisfactory differentiation of a particular species with other species in that genus. Mindful of this, 25 strains of *E. coli* were examined, and they showed distinct differences when cross-correlated with four selective *E. coli* strains. Same strain correlation provided high similarity coefficients, and that of different strains yielded relatively low similarity coefficients. The cross-correlation technique was used to monitor the general change in mass spectra during the growth and stationary phase of a bacterial culture (Arnold et al. 1999). Successive spectra provided high or low correlation values. A single plot of the correlation values provided a dynamic impression of the mass spectral changes over a 50 h growth period for *E. coli* cultures.

The general concept of analyzing many strains of a particular bacterium for classification and identification purposes was continued for a very different application. Twenty-three isolates of *Bacillus pumilus* were examined from different locations in spacecraft assembly facilities, locations in the Mars Odyssey spacecraft, and the International Space Station by Dickinson et al. (2004). In addition, MALDI mass spectral data were compared to the results of 16S ribosomal DNA (rDNA) sequence analysis and DNA–DNA hybridization, and Biolog bacterial enzyme analyses. Cross-correlation was performed on the *B. pumilus* isolates and 10 other *Bacillus* species to test for their differentiation. Twenty replicate mass spectra were obtained for each organism. The 18 *B. pumilus* strains produced relatively high correlation values with the *B. pumilus* ATCC 7061[T] isolate where all but two isolates achieved >0.62 correlation values. Correlation values of 0–0.48 were obtained for the *B. pumilus* isolates and the 10 other *Bacillus* species. The Biolog metabolic fingerprinting assays produced a relatively low identification analysis for the *B. pumilus* isolates, and the 16S rDNA tests did not perform as well as the DNA–DNA hybridization method. MALDI-MS provided similar information with respect to the DNA–DNA hybridization tests. Both methods segregated the *B. pumilus* isolates into two separate groups, and the linear correlation of the mass spectra correctly classified 14 of 16 *B. pumilus* isolates.

Automated extraction of selected masses was a central goal as reported by the K. Wahl group (Jarman et al. 1999; Wahl et al. 2002; Valentine et al. 2005). Sixty MALDI mass spectra, obtained over different days from the same sample, provided stability in the choice of masses selected to represent a bacterial genus, species, and strain in the database library. Further, the database was represented by mass and intensity with respective standard deviation values to compare against unknown or submitted spectra. This technique was successfully extended to 50 mixtures consisting of 2–4 bacteria (Wahl et al. 2002) in double-blind experiments performed on different days. Five replicate spectra were obtained for each sample mixture.

Cluster analysis results presented as dendrograms have been used to distinguish MALDI mass spectra from many different bacteria. These include studies of *Heliobacter* strains (Owen et al. 1999), *Enterobacteriaceae* including 11 *E. coli* strains (Conway et al. 2001), 28 isolates of *Staphylococcus* (Walker et al. 2002), a host of gram-positive and gram-negative bacteria (Wahl et al. 2002), 8 cultures of Mycobacteria (Hettick et al. 2004), and the separation of smooth and rough cell surface *Peptostreptococcus micros* (Anderson et al. 1999) preparations.

A group led by C. Fenselau published a number of papers on data analysis of bacteria in a systematic, refined fashion. Initially, MALDI mass spectral masses were compared to online bacterial protein databases without application of filter or weighting factors (Demirev et al. 1999). *B. subtilis* and *E. coli* were initially investigated with successful matching results. However, since their sheer numbers of masses in online databases are greater than that of most other bacteria, the statistics were skewed to the more densely populated bacterial protein databases such as *B. subtilis* and *E. coli*. This method of bacterial matching, however, may be independent of reproducibility issues since an experimental set of masses may be found in a comprehensive protein mass database (Demirev et al. 1999). This work was refined (Pineda et al. 2000) where the density of masses per unit mass interval was considered and investigations concentrated on the *E. coli* and *B. subtilis* organism databases. The *p*-value analyses estimated the probability of bacterial misidentification due to an accidental match between a set of experimental mass peaks and database proteins of an unrelated microorganism. The lower the *p*-value, the less likely the bacterial match occurred by chance, and 10–30 peaks characterized the experimental datasets. In this analysis, the possibility existed that the relatively low number of peaks compared to the hundreds of masses in a bacterial database may not yield robust statistics (Pineda et al. 2003).

Pineda et al. (2003) used experimental conditions in such a way that samples were enriched in specific components of bacteria that are ribosomal proteins, which were predominately extracted and observed in the MALDI mass spectrum as intact ribosomal protein masses. The *p*-values were used to test the significance of matching experimental spectra to online bacterial protein databases. Positive- and negative-mode MALDI analyses were investigated for *B. subtilis*, *E. coli*, *P. aeruginosa*, *Haemophilus influenzae*, and *Bacillus stearothermophilus*. Essentially, 100% correct identification of these organisms was produced in the interrogation of protein databases consisting of 38 organisms when at least 20 experimental masses per organism were obtained and searched. Model statistical considerations predicted a high rate of correct identification of the five organisms when compared to 1000 bacterial databases in the library.

Mass density considerations coupled with accurate mass assignments (Demirev et al. 2001a,b) were shown to provide a significant and impressive identification of *Heliobacter pylori* when searched in online, bacterial protein databases. The *H. pylori* 26995 sample had a significance value of 0.036 (low probability of a chance match), while the J99 strain had the next best match at a value of 0.065. Note that *H. pylori* 26995 and J99 only have 443 and 291 masses listed in their online protein databases. Most impressive were the relatively high probability misidentification values (parentheses) of *B. subtilis* (0.816), *Mycobacterium tuberculosis* (0.990), and *E. coli* (0.998) that, respectively, contain 1420, 1058, and 2030 online protein database masses. These numbers of database masses are significantly greater than that of *H. pylori*, yet the algorithm parameters were able to target the correct *H. pylori* 26995 strain compared to the J99 strain database.

Wang et al. (2002a,b) provided a fundamentally different treatise for MALDI mass spectral bacterial characterization in that an in house–generated database was shown to provide better matching statistics than online databases. Many experimental masses of bacteria usually are not found in the respective online database. *E. coli*, *Bacillus megaterium*, and *Citrobacter freundii* were used to test this hypothesis, and there were 2997, 55, and 47 protein mass entries, respectively, in the 2–20 kDa range in the public online databases. As expected, the experimental MALDI masses from all three bacteria produced *E. coli* as the top match when interrogated with online protein databases. In house–derived experimental mass tables provided significantly better mass-matching performance with bacterial MALDI mass spectra compared to that of online databases. This is especially true for online bacterial databases containing low numbers of mass entries. A benefit of in house–generated databases is that they include masses that constitute PTM moieties on the proteins. This is inherent in in-house databases while it is essentially absent in public proteome databases. Another benefit is that experimental, in-house databases include protein masses that are actually expressed at levels detected by MS as opposed to online mass entries of proteins that are expressed in undetectable, very low, moderate, and very high numbers of copies in a bacterial cell.

M. tuberculosis is a pathogen, and there exists a significant number of nonpathogenic *Mycobacterium* species. Six species including *M. tuberculosis* were investigated by MALDI-MS, and their eight replicate spectra were transformed into multivariate data space (Hettick et al. 2004). Discriminant and canonical variate data analyses were performed in order to differentiate among the six mycobacterial species. Upon sample optimization, a plot of the first three canonical variate dimensions described 85% of the total variance in the dataset, and all six species were easily differentiated despite the complexity of the raw spectral data. Seventy-seven masses were used as input to the canonical variate analysis. From this analysis, certain masses appeared to act as biomarker ions for five of the six *Mycobacteria* species. *Mycobacterium* species *fortuitum*, *kansasii*, *intracellulare*, *tuberculosis*, and *avium* displayed unique ions while *Mycobacterium bovis* BCG provided no unique ions. However, the latter could be differentiated by the intensity distribution of the mass spectral ions.

Recently, Tao et al. (2004) provided an interesting method for the differentiation of microorganisms by using a database of biomarker masses. A database was constructed by MALDI-MS analysis of a bacterial culture between 9 and 12 selected times during a growth period between 8 and 48 h. Ten different organisms were used to create a database. Masses were tabulated for each growth time, and each mass was noted as to when it appeared in each different growth time. Some masses appeared at every growth time in the exponential and stationary growth phases and other masses appeared in only one or a few growth times. If an experimental mass is found in one or more of the database organisms, then the weight factor for that mass is annotated under each organism. All weight values for each mass match are summed for each database organism and the highest sum of the weight values determines the identity of the sample.

Although vegetative bacteria produce some different proteins when they are cultured in different growth media, positive identification with MALDI-TOF-MS is still possible with the protocol established at the Pacific Northwest National Laboratory (Jarman et al. 2000). A core set of small proteins remain constant under at least four different culture media conditions and blood agar plates, including minimal medium M9, rich media, tryptic soy broth (TSB), or Luria-Bertani (LB) broth, and blood agar plates, such that analysis of the intact cells by MALDI-MS allows for consistent identification (Valentine et al. 2005).

MALDI-MS in combination with unsupervised pattern recognition algorithms such as hierarchical cluster analysis or principal component analysis (PCA), or supervised algorithms such as artificial neural networks (ANN) has shown mixed degrees of success for analyzing microbial mass spectral data. Statistical studies of bacterial MALDI-MS experiments have provided some insight into the factors reducing the success of these approaches, showing that, whereas some mass spectral peaks are highly reproducible and appear consistently, others appear much less reliably. Two main sources of variability can be identified in microbial MALDI-MS experiments. The first originates in changes in culture conditions that produce changes in protein expression levels, altering the intensity and/or occurrence of the observed mass spectral peaks. It is well known that culture conditions have to be kept as constant as possible to ensure reproducibility of the obtained MALDI fingerprints. In cases where culture conditions change, as when different media batches are used, correction algorithms can be applied to transform the new set of fingerprints with varying degrees of success. A second source of variability in the MALDI data originates in the intrinsic reproducibility of the MALDI processes, including variables such as the sample preparation protocol, the type and quality of matrix chosen, ionization suppression effects, mass scale drifts, and the possible impact of automatic data acquisition algorithms. In an effort to standardize the conditions for MALDI bacterial fingerprinting, Wunschel et al. (2005) have recently studied the sources of bacterial MALDI mass spectral variability in a comprehensive inter-laboratory study. Soft modeling methods that create optimal linear relationships among constructs specified by a conceptual model, like PCA and PLS, can successfully mitigate the detrimental effects of noisy and highly collinear spectra, such as those found in bacterial MALDI data. Because PCA relies on the generation of scores on orthogonal principal components, it attempts to capture the directions of maximum variance, not seeking to capture "among-group" and "within-group" *differences* of the investigated objects. Soft modeling by partial least squares discriminant analysis (PLS-DA) is a more recent *supervised* pattern recognition approach that attempts to overcome some of the drawbacks observed in PCA. During PLS-DA, the principal components are rotated to generate latent variables (LVs), which maximize the discriminant power between different classes, not the total mass spectral variance as in PCA and, therefore, class separation is greatly improved. For example, Pierce et al. (2007) presented results on the identification of *Coxiella burnetii* cultures using PLS-DA of MALDI-TOF mass spectral peaks for a whole cell. The combination of data smoothing, denoising, and binarization with PLS-DA multivariate analysis allowed differentiating seven *C. burnetii* strains in a training set containing spectral data obtained on four different days within a period of 6 months. In addition, they performed a two-class discrimination of *C. burnetii* phase I strains versus phase II strains to assess the antigenicity of a given culture. All models were validated by classifying unknown *C. burnetii* samples run on a fifth day.

The importance of keeping constant experimental conditions and their influence on the reproducibility and discriminative power of MALDI-MS methods is demonstrated in Figure 3.1, which shows mass spectra of *B. anthracis* obtained with different sample solvents. A combination of 0.1% trifluoroacetic acid (TFA) and chloroform–methanol (1:1) (solvents I and II) resulted in the best signal for *B. anthracis*. Although many common peaks were present in the five spectra when different sample solvents were used, the peak numbers, the relative intensities of peaks, and the *m/z* ranges were different. Similar effects on the spectra of those sample solvents were also observed when other bacterial samples were examined (Liu et al. 2007).

Recently, on the basis of numerous results, Liu et al. (2007) proposed a universal sample preparation method for MALDI-TOF-MS of bacteria. In order to test if the same protocol is applicable to more bacterial species with different characteristics, *Staphylococcus aureus* 658, *Burkholderia cepacia* 855, and *E. coli* JM109 were also analyzed (Figure 3.2). Of the five bacterial species tested, *B. anthracis* is gram positive and spore producing; *S. aureus* is gram positive and nonspore producing; *E. coli*, *Yersinia pestis*, and *B. cepacia* are gram negative; and *B. cepacia* has high extracellular polysaccharide content. It was found that peaks with different *m/z* values could readily distinguish these five species from each other and all of their mass spectra were represented by more than 20 *m/z* values with high sensitivity (Figure 3.2).

3.3 WHOLE-CELL CHARACTERIZATION THROUGH MALDI-FTMS

Mass spectra obtained from whole cells of microorganisms through the use of MALDI-MS represent *m/z* observations that are not assigned as specific proteins or lipids because such identifications are not possible with low-resolution mass data. However, MALDI FTMS that provides high-resolution mass separations of ions makes accurate assignments of molecular identities possible. Thus a significant improvement over fingerprint-type approaches can be achieved based on accurate mass and isotopic abundance values obtained from each experiment. Nevertheless, sorting through existing databases in order to use them with high-resolution FTMS data represents a challenging task.

Spectral peaks located in the low-molecular-weight *m/z* segment of the mass spectrum usually represent compounds derived from lipid components like glycerides, phospholipids, glycolipids, or protein fragments. On the other hand, the high mass segment of the spectrum generally represents proteins. Due to the huge number of peaks observed in the spectrum, data analysis is quite complicated. However, examples of successful approaches exist. For instance, Jones et al. (2003) used a

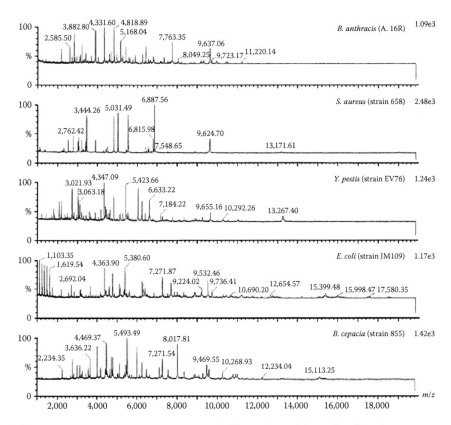

FIGURE 3.2 MALDI-TOF-MS spectra of different bacterial species. Samples were pre-pared under the same experimental conditions, i.e., bacteria were grown under standard con-ditions and then treated with solvents I and II. CHCA in matrix solvent A was used as the matrix and the dried droplet method was used. (From Liu, H. et al., *Appl. Environ. Microbiol.*, 73(6), 1899, 2007.)

set of logical rules based on lipid chemistry, simple mathematical formulas, and mass defect plots to assign almost all peaks in this part of a bacterial spectrum. It seems that intact cell MALDI-FTMS of bacterial lipids could complement the use of proteomics profiling by MS because it relies on accurate mass measurements of cel-lular components that are relatively stable in comparison with proteins. Furthermore, Jones et al. (2003) showed that in addition to information regarding *E. coli* lipids in the low mass region, the accurate mass MALDI-FTMS could be used to characterize specific ribosomal proteins directly from *E. coli* cells. This was accomplished directly from whole cells without fractionation or concentration. Moreover, these spectra also provided information regarding *E. coli* lipids in the low mass region (Jones et al. 2003).

In summary, high-resolution mass spectra obtained with FTMS from whole cells showed significant improvement in apparent mass-resolving power and mass mea-surement accuracy. Moreover, the chemical interference was significantly reduced.

As a consequence, it becomes possible to detect subtle details in the chemistry of the organism, such as the presence of both PTM and unmodified versions of the same proteins (Stump et al. 2003).

3.4 SELDI

SELDI is a modification of a MALDI technology that is used in combination with TOF-MS for analysis of proteins from complex mixtures. SELDI uses chip arrays as targets that are coated with different chromatographic surfaces to achieve selective capture of protein subsets for further analysis by TOF-MS. Protein expression profiles give a fingerprint pattern for the designated group of samples that can be used for diagnostic purposes. This technology was used for analysis of specific protein profiles of bioagents as well as for analysis of host responses through profiling of plasma proteins to detect biomarkers of a viral infection. For example, Poon et al. (2004) applied this technology to viral disease outbreaks such as severe acute respiratory syndrome (SARS). It was concluded that proteomic profiling of plasma proteins could form a basis for a serum profiling assay suitable for the early detection of SARS infection and distinguishing SARS from influenza and other causes of fever.

Dahouk et al. (2006) analyzed fresh whole-cell lysates of *Brucella abortus*, *Yersinia enterocolitica*, *Vibrio cholerae*, and *Francisella tularensis* using four different chromatographic surfaces [strong anion exchange (SAX2), weak cation exchange (WCX2), immobilized copper cations (IMAC-Cu), and a reverse phase (H50)]. Protein chips were equilibrated with suitable binding buffers and incubated with samples followed by a short wash and the addition of saturated sinapinic acid as a matrix. Proteins were laser desorbed and mass analyzed in the range from 3000 to 20,000 Da. The different chromatographic surfaces captured 170 proteins within the specified mass range and the authors demonstrated that SELDI profiles complemented 2D patterns revealed by gel electrophoresis, especially well in the low-molecular-weight range. Moreover, all bacterial species revealed significantly different and highly reproducible SELDI spectra that allow a fast and clear differentiation of the bacteria (Dahouk et al. 2006).

Lundquist et al. (2005) demonstrated that SELDI technology could be used to discriminate *F. tularensis* up to the subspecies level that with traditional methods is difficult and tedious. They added bacterial thermolysates to anionic, cationic, and copper ion–immobilized affinity chip surfaces and used SELDI-TOF-MS protein profiles to characterize each bacterial strain. The cluster and PCA results of data for the investigated *Francisella* strains grouped them according to the recognized subspecies. In addition, PLS-DA of the protein profiles also identified proteins that differed between the strains. Thus, protein profiling of bacterial lysates with SELDI-TOF-MS holds promise for rapid phenotypic identification of bacteria.

3.5 PY-MS

Pyrolysis indicates decomposition caused by thermal energy at the relatively high temperature. During this process, dissociation of covalent bonds and their rearrangements brought about by heat produces pyrolyzates of microbial biomolecules.

These pyrolyzates may be analyzed directly by MS (Py-MS) or after pre-separation of pyrolyzate constituents through GC (Py-GC-MS). It is assumed that under appropriate conditions, thermal degradation products reflect to a large extent the original structure of pyrolyzed materials and for decades pyrolysis was the only sample pretreatment method for MS analysis of solids (for review, see Dworzanski and Meuzellar 1988).

Py-MS was commonly applied for bacterial fingerprinting, following the first reports published in the 1960s on the applicability of analytical pyrolysis techniques to clinical and pharmaceutical microbiology. The important advantages of Py-MS techniques include the small sample size required, the high sample throughput, and the compatibility of the data with computerized evaluation that allowed a seamless statistical analysis of multivariate data generated using Py-MS. However, both instrumental and biological factors influence the outcome of the analysis. Therefore, standardization of instrumental conditions is essential to achieve an acceptable level of long-term and inter-laboratory reproducibility while standardized culturing and sampling are required to minimize environmental causes of biological heterogeneity among the samples. The characterization of biological materials and biothreat agents by MS analysis of pyrolyzates obtained using diverse pyrolysis techniques was extensively reviewed by Snyder (1992). More recent applications of this technology were reviewed by Timmins and Goodacre (2006).

Pyrolysis of bacteria produces decomposition products, allowing chemical and biochemical inferences to be made about cellular components of a sample. The proposed origin of identified thermal degradation products observed during whole-cell bacterial pyrolysis is usually validated by comparison with pyrolytic products generated from suitable model compounds or cell fractions pyrolyzed under identical conditions (Meuzalaar et al. 1982). Pyrolysis products of cytosol components such as nucleic acids and proteins are composed of very similar compounds, and although used for pyrolytic typing of different strains, thermally generated products of these molecules do not provide enough biomarker information to distinguish bacterial species or even to differentiate between prokaryotic and eukaryotic organisms. However, pyrolysis products derived from carbohydrates, lipids, and other components, e.g., dipicolinic (Snyder et al. 1996) or poly(3-hydroxyalkanoic) acids (Watt et al. 1991), have been widely utilized for bacterial discrimination due to their substantial discriminative power. Furthermore, a significant reservoir of taxonomic information can be revealed by structural and compositional analyses of bacterial cell envelopes. For instance, bacterial cell walls are composed of unique biopolymer peptidoglycan, which forms a highly cross-linked network resembling a rigid, insoluble skeleton that determines the shape of the bacterial cells (Höltje 1998). However, gram-positive bacteria contain a 20–80 nm thick cell wall that forms a layer (murein) around the cytoplasmic membrane, while the peptidoglycan layer in gram-negative bacteria is much thinner (2–5 nm) and is surrounded by a second membrane rich in lipopolysaccharide (LPS) biopolymers. Hudson et al. (1982) applied Py-GC-MS to the gram typing of bacteria using levels of both acetamide and propionamide, which originates from the pyrolysis of the lactyl residue in a muramic acid component of peptidoglycan. Recently, Dworzanski et al. (2005) reported that the Py-GC module of a fielded biodetector generates a set of specific products that reflect the biochemical composition of bacterial cell envelopes. A prominent pyrolysis biomarker derived from the cell walls of gram-positive bacteria has been identified as pyridine-2-carboxamide that originates from the thermal rearrangements of the *meso*-diaminopimelic acid and/or L-lysine peptide residue cross-linker species in peptidoglycan. Biomarkers produced by pyrolysis of gram-negative organisms were found to originate from the lipid A portion of LPS that are localized on the cell surface. These biomarkers include pyrolysis products of the 3-hydroxymyristate fatty acid residues such as 1-tridecene, dodecanal, and methylundecylketone. Nitrile derivatives from amide-bound fatty acids and free fatty acids originally present as ester-bound moieties in lipid A are also observed upon bacterial pyrolysis (Snyder et al. 2005).

Analysis of complex samples is usually associated with a lengthy and labor-intensive sample preparation task. Hence pyrolysis is an attractive sample-processing technique, because it provides simplicity, rapidity, and relative ease of integration with chromatography- and spectrometry-based analytical methods. Pyrolysis reactors have been applied as sample-processing modules for outdoor, fielded continuous monitoring of bioaerosol particulates with respect to biodefense purposes (Snyder et al. 1996, 1999, 2000; Dworzanski et al. 1997).

Very fine methods developed for the identification of bacterial cultures using Py-MS and Py-GC-MS provide reproducible mass spectra that proved to be characteristic of species and even particular strains of microorganism. In spite of their highly reproducible nature, most molecular information is lost in these experiments. Nevertheless, pyrolysis is still used in some laboratories for discrimination of species.

3.5.1 Pyrolysis-MAB-MS for Rapid Bacteria Identification and Subtyping

Researchers from the National Center for Toxicological Research (FDA) developed a method for rapid phenotypic characterization of bacteria using a unique ionization method for generating pyrolysis mass spectra. In this method, highly reproducible pyrolysis fragmentation patterns are generated due to narrow energy distribution obtained during MAB of pyrolyzates (Faubert et al. 1993). The MAB ionization is achieved by using neutral gas atoms with the elevated internal energy obtained by passing gas molecules through high voltage–induced plasma. The metastable atoms are directed on the pyrolyzate and transfer of energy occurs. The MAB gas provides the way of obtaining pyrolysis spectrum blind to sample contaminants that are usually present in high ionization energy spectra. Furthermore, the ionization energy is user selectable and can be optimized for a specific group of compounds.

Wilkes et al. (2005a) compared Py-MAB-MS spectra of *Salmonella* strains to their serovars, pulsed-field gel electrophoresis patterns (PFGE), antibiotic-resistance profiles, and minimum inhibitory concentration values for diverse antibiotics. They demonstrated that DA of spectral patterns allows distinguishing *Salmonella* strains by serovar (97% correct) and PFGE groups. They also reported that Py-MS with MAB ionization and pattern recognition appeared suitable for rapid infra-specific comparison of *Vibrio* isolates and concluded that the integration of this analytical system with advanced statistical analysis should be examined further for potential utility in clinical and public health diagnostic contexts (Wilkes et al. 2005b).

It is interesting to note that this technique was also investigated as a rapid tool (data acquisition took 7 min per sample) to distinguish potential bioterror hoax materials from samples containing pathogenic agents. Several materials containing strains of *Vibrio parahaemolyticus* (one produced the tdh toxin), *Salmonella enterica* serotypes, spores of *B. thuringiensis*, and potential hoax materials, such as flour, starch, methyl cellulose, and xanthan gum, were analyzed (Wilkes et al. 2006). Pattern analysis of Py-MAB-MS spectra distinguished bacterial samples from hoax materials and differentiated *B. thuringiensis* spores, *Vibrio*, and *Salmonella* strains, thus suggesting that this technique may differentiate bioterror agents from potential hoax materials.

3.6 FATTY ACID PROFILES

3.6.1 GC-MS FOR ANALYSIS OF FATTY ACID METHYL ESTER PROFILES

Analysis of microbial fatty acids by means of GC–liquid chromatography (LC) has been commonly carried out for the characterization, identification, and differentiation of bacterial strains during last few decades (Jantzen 1984; Fox and Morgan 1985). Compositions of cellular fatty acids, as accumulated in the literature published during last decades, are also commercially available as computer databases (Microbial Identification System, MIDI, Newark, DE) and applications of fatty acid profiling for microbial taxonomy are presented in *Bergey's Manual of Systematic Bacteriology*. The system developed by MIDI converts cellular fatty acids from pure cultures of bacteria to fatty acid methyl esters (FAMEs) and uses a GC for separation and identification purposes. Samples are harvested, saponified, methylated, and extracted into an organic phase for analysis. Suitable pattern recognition software is used for matching a FAME profile with database for identification of isolates. This method has been used to identify and differentiate *Bacillus* spores, *Burkholderia* spp. (Inglis et al. 2003), *Francisella* spp., and *Yersinia* spp. (Leclercq et al. 2000). Recently, MIDI Inc. introduced the Sherlock Bioterrorism Library that can be added to its identification system to specifically target biothreat agents and challenge organisms. The MIDI Sherlock system containing the MIDI BIOTER database (ver. 2.0) has been awarded AOAC Official Methods of Analysis status for confirmatory identification of *B. anthracis* (AOAC International 2004).

This biochemical method is inherently time-consuming as it requires pure isolates for the tests to be accurate. In addition, well-trained, experienced technicians are required to accurately identify and subsequently handle further testing that may be necessary. Although this system is more convenient than manual procedures, it typically cannot be used in the field and its usefulness may be limited by the extent of their databases, which must be updated on a regular basis.

To overcome this slow and laborious sample preparation procedure, Dworzanski et al. (1988, 1990) developed a pyrolytic in situ methylation technique (Figure 3.3) that opened the possibility for rapid bacterial detection and identification using field-able instruments. This technique was implemented in the CBMS Block II system, however, without any chromatographic separation step that was used in the original

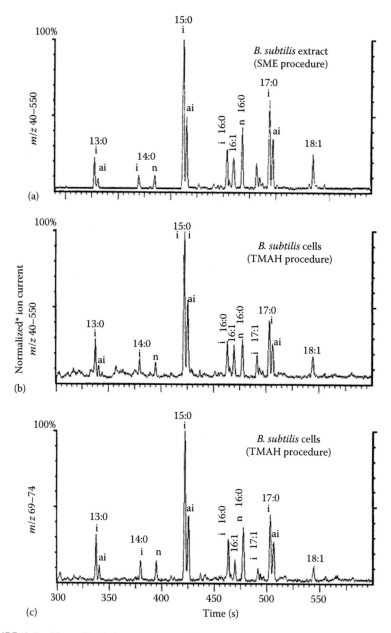

FIGURE 3.3 Normalized ion current profiles of *B. subtilis* ATCC 6633. (a) Total ion chromatogram of the extract obtained by the saponification–methylation–extraction (SME) procedure and (b) total ion chromatogram of pyrolytic products of whole cells obtained by the TMAH procedure; and (c) reconstructed mass chromatogram (*m/z* 69–74) from the total ion current chromatogram shown in panel b. Peak designation: Number before the colon refers to the number of carbon atoms of fatty acid and the number after the colon is the number of double bonds; i, iso; ai, anteiso; n, normal (unbranched) acid. (From Dworzanski, J.P. et al., *Appl. Environ. Microbiol.*, 56(6), 1717, 1990.)

invention. Therefore, the system lost the capability to distinguish isomers of fatty acids, and the diminished resolving power made it impossible to identify bacteria on lower taxonomic levels.

3.6.2 Thermolysis/Methylation–Ethanol Chemical Ionization–QIT for Detection

Thermolysis or pyrolysis of biological warfare agents gives many products that can be uniquely traced to source components, that is, cellular molecules like lipids, proteins, and nucleic acids, and used for the detection and identification purposes based on analysis of characteristic fingerprints. The previous fieldable system based on the implementation of this principle is known as CBMS Block I. In the CBMS Block I approach, concentrated bioaerosol particles are deposited on a quartz filter in a flow-through tube, pyrolyzed, and the pyrolysis products transferred to a MS through a silicon membrane vacuum interface. Inside an ion trap, these components are ionized using an electron impact and mass analyzed. The obtained mass spectra were used as fingerprints of biomarkers and were used for classification of sampled agents such as bacterial, viral, or toxin, but without the capability to identify a specific agent.

The currently fielded CBMS Block II, a component of the Biological Integrated Detection System (BIDS), also uses a fingerprint approach. However, this system rapidly pyrolyzes cellular components from a collected aerosol in the presence of a methylating agent, thus producing methylated derivatives of fragmented cellular constituents. They are ionized using ethanol CI and mass analyzed with an ion trap MS to generate mass spectral fingerprints.

In the upgraded version of this approach (CBMS II), the pyrolyzer assembly is coupled with the tetramethylammonium hydroxide (TMAH) solution delivery system required for derivatization of polar components. Furthermore, the availability of high-capacity turbomolecular pumps allowed replacing the inefficient membrane with capillary inlet module, and the application of a CI of analytes with ethanol. The use of ethanol CI as a CI reagent provides additional selectivity due to higher proton affinity in comparison to hydrocarbons.

In this modified mode of operation, a mixture of particles and methanolic TMAH undergoes thermolysis that produces methyl esters of cellular fatty acids (Figure 3.4). The reduced polarity of derivatized fatty acids combined with the capillary interface improves the efficiency of biomarker transport, thus increasing the overall sensitivity (Barshick et al. 1999).

The ability to differentiate microorganisms using pyrolysis CI ion trap MS is demonstrated in Figure 3.5. In this graph, a PCA was applied for strain discrimination. Based on total ion profiles, the ability to distinguish the bacteria was limited to the genus level using electron ionization but included the subspecies level using chemical ionization.

Overall, the identification of biological agents through fatty acid analysis, especially by analysis of molecular ions without further fragmentation, precludes distinguishing positional isomers and is not sufficiently robust for the identification purposes. Furthermore, the nature of the fatty acids in bacterial cultures is known to vary with growth conditions such as temperature.

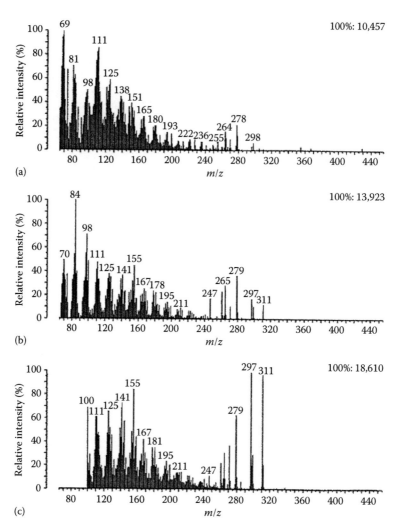

FIGURE 3.4 Average mass spectra for pyrolytic methylation of *Brucella melitensis*. Bacterial profiles are shown for (a) electron ionization, (b) water chemical ionization, and (c) ethanol chemical ionization. Mass spectra were averaged across 50 scans. (From Barshick, S.A. et al., *Anal. Chem.*, 71(3), 633, 1999.)

While potentially a useful rapid screening tool for biological agents on the battlefield, it is unlikely that this system will have the specificity needed for the extremely low false alarm rates required of a rapid identifier in a detect-to-warn architecture. The pyrolysis fragmentation pattern is a function of many variables, and it has not been demonstrated that these signatures offer sufficient discrimination in a complex background to support a 10^{-6} false alarm rate. Hence, advances in genomic/proteomic methods may make those approaches a more viable alternative to fatty acid analysis.

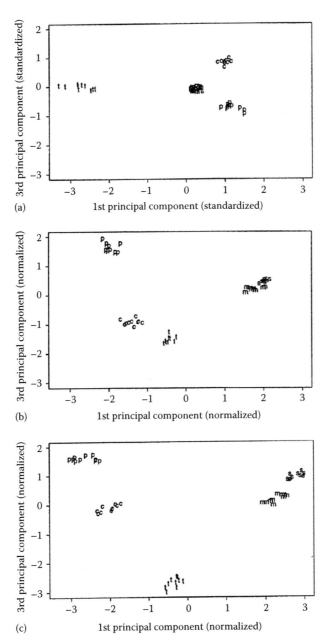

FIGURE 3.5 PCA plots showing first and third principal components for (a) electron ionization, (b) water CI, and (c) methanol CI analysis of five bacterial samples in total ion space. The letters correspond to the species name of each bacteria: (s) *Brucella suis*, (m) *B. melitensis*, (p) *Y. pestis*, (c) *V. cholerae*, and (t) *F. tularensis*. The data represent 10 replicate analyses of each bacterial sample.

3.7   DIRECT INFUSION ESI-MS OF CRUDE CELL EXTRACTS

ESI-MS is usually applied to cell components pre-fractionated using diverse chromatographic or electrophoretic techniques. However, it was demonstrated that by using a direct infusion of bacterial cell extracts, that is, without separation steps, it was possible to identify important bacterial markers like phospholipids (Smith et al. 1995), glycolipids (Wang et al. 1999a,b), or muramic acid (Black et al. 1994). Furthermore, even a direct infusion of whole bacterial cells was investigated for rapid discrimination of microbes and the ability of ESI-MS to provide information-rich spectra from cell suspensions was reported by Goodacre et al. (1999). This approach was subsequently used to discriminate aerobic endospore-forming bacteria and demonstrated the feasibility of measuring liquid samples with minimal sample preparation that can be useful for discrimination at the subspecies level (Vaidyananathan et al. 2001). The analyses were performed using bacterial cells suspended in aqueous acidic acetonitrile that was infused with a syringe pump and electrosprayed into a MS. Multivariable data analysis, that included PCA of raw spectra, allowed for species-level discrimination of investigated samples. It was achieved by discriminant function analysis of principal component scores and cluster analysis of Euclidean distances of discovered bacterial species clusters. The relatedness among analyzed samples was similar to results obtained from sequence analysis of small ribosomal subunit (16S rDNA) or classical biochemical tests. Unfortunately, this mode of operation is associated with the unacceptably high level of contaminants introduced into an MS system, thus making this approach impractical for field operations.

3.8   DESI-MS

DESI coupled with MS provides sensitive and useful information by performing direct analysis, e.g., on growing cultures in Petri dishes (Takats et al. 2005). Recently, a few research groups evaluated this novel ionization technology for the potential detection and discrimination of bacteria using MS detectors. Inventors of this technology from the Purdue University reported a successful discrimination of five *E. coli* and *Salmonella typhimurium* strains (Song et al. 2007). Bacteria were examined as suspensions of freshly harvested cells deposited onto a glass slide over an area about 0.5 cm^2. Positive ion DESI-MS spectra of bacteria were recorded with a linear ion trap after evaporation of water, and analysis performed by directing electrosprayed solvent onto a glass slide (Figure 3.6). PCA was performed on the spectral data and results obtained are presented as a score plot of the first two principal components in Figure 3.7. The score plot indicates that five strains of closely related bacteria could be well separated, which corresponds to the differentiation between individual strains.

The usefulness of DESI-MS for the detection, characterization, and differentiation of seven bacterial species was also reported by researchers from the University of Wyoming. The authors explored the same day reproducibility of analyses as well as the influence of growth conditions on the results of long-term reproducibility evaluated using a PCA technique (Meetani et al. 2007). They concluded that DESI mass spectral profiles of bacteria depend heavily on the experimental design; therefore,

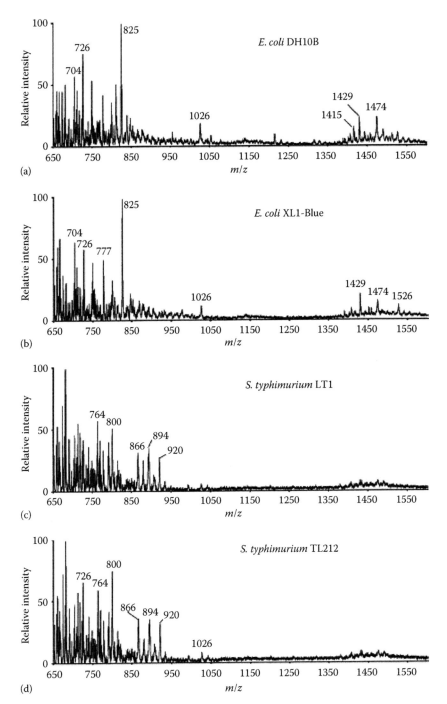

FIGURE 3.6 Typical DESI mass spectra of (a) *E. coli* DH10B, (b) *E. coli* XL1-Blue, (c) *S. typhimurium* LT1, and (d) *S. typhimurium* TL212. (From Song, Y. et al., *Chem. Commun. (Camb.)*, 7, 61, 2007.)

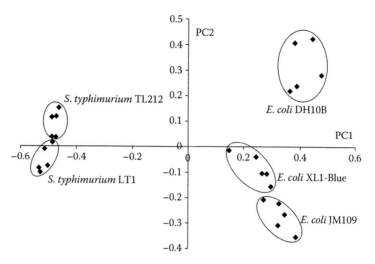

FIGURE 3.7 PCA score plot of five strains of bacteria cultivated and analyzed under identical conditions. (From Song, Y. et al., *Chem. Commun. (Camb.)*, 7, 61, 2007.)

profiling of bacteria should be performed by using isolates cultured under the same standardized conditions or the analysis should be focused on specific biomarkers.

3.9   BIOAEROSOL MS

A rapid, reagentless online analytical technique to sample and detect bioaerosols, at the individual particle level of resolution, called bioaerosol MS (BAMS), is being developed at Lawrence Livermore National Laboratory for national security and public health applications that require detection of airborne bioagents such as vegetative bacteria, bacterial spores, viruses, and biological toxins (Steele et al. 2003; Fergenson et al. 2004; Russell et al. 2004). Although the size and cost of BAMS is currently greater than some other techniques, it will be useful where real-time, unmanned, continuous operation is required and development of a smaller and less-expensive version is under way. The BAMS technique is based on single-particle aerosol MS, which has been applied to the analysis of a range of biological particles (Gieray et al. 1997).

In BAMS, many individual airborne particles are sampled per second directly from the air and accelerated to size-dependent speeds in a vacuum system, where the individual particles are then sized by the time delay measured between two laser scattering events. Each particle in the airflow is interrogated using laser-induced fluorescence, and particles having fluorescent properties of characteristic for pathogens are selected and broken into their components using a laser desorption ionization. The molecules are desorbed and ionized using high-power 266 nm laser light, and the fragments are chemically analyzed by using dual-polarity TOF-MS. Note that sequential sorting of particles ahead of analysis by TOF-MS prevents flooding the MS instrument, while all ions, positive and negative, are analyzed in parallel. The mass spectral signatures acquired (Figure 3.8) are compared with those of known target particles stored in a library which is interrogated for matches.

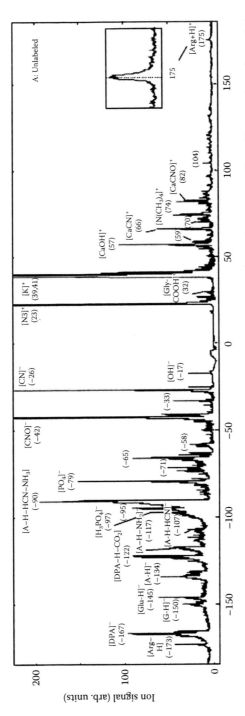

FIGURE 3.8 Dual-polarity average mass spectrum of *Bacillus atrophaeus* (324 spectra) grown in bioexpress medium. The inset in each panel shows the protonated arginine peak.

BAMS has been shown to be able to pick out features in the mass spectra that clearly distinguish between closely related species. For example, BAMS has already been shown to be a promising technique for the detection of *Bacillus* spores of interest to biodefense applications, and the differentiation of two *Bacillus* spore species has been demonstrated (Fergenson et al. 2004). BAMS is currently being developed further as an analytical method for the rapid, noninvasive diagnostics of infectious disease in respiratory effluent. Such a method would permit a more rapid response to public health concerns associated with infectious diseases like tuberculosis and influenza. Requiring no reagents, the instrument needs only weekly maintenance and it can be operated by remote control. BAMS is inherently a very fast monitoring device with detect-to-warn times <1 min on the order of seconds if the highest sensitivity is not required. Overall, the speed of BAMS ensures a rapid warning of pathogens introduced into the atmosphere. The prototype instrument is undergoing "ruggedization" in order to demonstrate that it is robust enough for field use.

3.10 DISCUSSION

The techniques used in this chapter have all made contributions in many areas, but have labored to be successful in the accurate identification of microbes. These methods represent the early limitations of the science available at the time and although useful for other applications often found difficult to be reliable when looking at complex biological mixtures containing multiple microbes of various types and concentrations.

4 Sequence Information Derived from Proteins or Nucleic Acids

Samir V. Deshpande

CONTENTS

4.1 INTRODUCTION

Microorganisms have traditionally been classified and identified by their physiological and biochemical properties or chemotaxonomic characteristics. However, recent advances in molecular biology suggest that the detection, classification, and identification of microorganisms reflecting relationships encoded in nucleotide sequences of nucleic acids or amino acid sequences of proteins are much more reliable.

4.2 PROTEINS

The classical approach for a global proteome analysis relies on protein separation by high-resolution 2D electrophoresis to obtain individual molecules for MS investigations (Tonella et al. 2001). In classical bottom-up methods, separated proteins are in-gel trypsinized, and the released peptides are identified by mass mapping or by analyzing product ion mass spectra obtained through CID or post-source decay (PSD, Chalmers and Gaskel 2000). Important technical advances related to 2D-PAGE and protein MS have increased the sensitivity, reproducibility, and throughput of proteome analysis. However, there are also important disadvantages to 2D-PAGE. On the one hand, the technology includes a bias against insoluble and high mass proteins or partial chemical degradation of basic proteins and on the other, it cannot be interfaced directly to a MS platform, thus making this method a laborious and time-consuming approach for routine applications.

The more recently developed shotgun approach uses a bottom-up strategy relying on a global, proteome-wide digestion of microbial proteins with proteolytic enzymes, followed by mass fingerprinting or microsequencing of peptides released from dominating proteins using MALDI-MS/MS technology (Section 4.2.1). However, substantial improvements in the scope of sequence coverage and reliability can be achieved through separation of peptides by LC or capillary electrophoresis (CE) prior to ESI-MS/MS analysis (Aebersold and Goodlett 2001; Wolters et al. 2001) as described in Sections 4.2.3 and 4.2.4. The application of this approach to microbial analysis has been substantially improved through the development of a novel method for analysis of sequence-to-bacterium (STB) assignments developed by Dworzanski et al. (2004a,b). In Section 4.2.7, a case study is presented as an example of MS-based classification of bacterial samples by using peptide-sequencing information revealed by high-throughput database searches.

Finally, the full characterization of a protein primary structure might benefit from emerging technologies based on a top-down MS approach. In this approach, an accurate measurement of relative molecular weight value for an intact protein is combined with direct dissociation of protein ions into fragment peptide ions, revealing amino acid sequence information. These pieces of information are combined to yield data suitable for the identification of proteins (Kelleher 2004; VerBerkmoes et al. 2004).

4.2.1 Peptide Mass Fingerprinting and MALDI-MS/MS of Peptides

When a few proteins are overexpressed in a bacterial cell, a separation step may be eliminated and the basic peptide mass fingerprinting (PMF) technique can be applied directly to the whole-cell protein digest. This approach takes advantage of the preferential ionization of peptide ions from the MALDI target. For MALDI-MS mass mapping analyses, either the intact bacterial cell or the protein extract can be subjected to protease (usually trypsin) digestion and analyzed without purification and pre-separation steps. Halden et al. (2005) presented a study on the PMF of gram-negative *Sphingomonas wittichii* strain RW1 that targeted a specific substrate-induced enzyme. Cells of *S. wittichii* were sonicated, and centrifuged protein supernatants were treated with trypsin followed by MALDI-MS analysis. PMF analysis was used to ascertain the presence of the targeted enzyme. In silico, peptide mass tables of protein subunits were generated and compared to experimental mass spectra. PMF analyses of the environmental cultures did not produce a match to the *S. wittichii* in silico database.

The PMF concept was augmented by Warscheid and Fenselau (2003), in which the family of small acid-soluble proteins (SASP) of *Bacillus* species was investigated. The on-probe digestion of spores from *Bacillus* spp. with immobilized trypsin cleaved the proteins into peptides. These peptides were used for microsequencing using tandem MS techniques and standard database searches. This in turn produced a table of identified SASPs that are proteins in the 6600–9200 Da range. The relatively limited set of SASPs could provide distinguishing capabilities for *B. cereus*, *B. thuringiensis*, *Bacillus subtilis*, *Bacillus globigii*, and *B. anthracis* Sterne. Mixtures of the bacilli could be distinguished from the unambiguous sequencing of selected peptides.

This was followed by the production of a custom-made database comparing the in silico trypsin digestion peptides from all SASPs contained in the online database of *Bacilli* and *Clostridia* organisms (English et al. 2003). Experimentally generated peptide sequences of extracted SASPs from bacterial samples were searched in the SASP database. Analysis of the *p*-values was used to test for false matches. *B. cereus* T, *B. thuringiensis* kurstaki, and *B. anthracis* Sterne were shown to have distinguishing peptides while *B. globigii* proved to have no discriminating peptides. This work used only a TOF-MS analysis. A quadrupole ion trap TOF system increased the resolving power of the precursor peptide masses (Warscheid et al. 2003) so that full online protein database searches provided distinguishing peptide features for all three bacteria in a mixture of *B. thuringiensis* kurstaki, *B. globigii*, and *B. subtilis* as well as in a mixture of *B. cereus*, *B. globigii*, and *B. subtilis*. The ion trap allows for a greater density of peptides to be collected before the TOF analysis and, as a result, an increased amount with more different types of peptides emerged.

4.2.2 Atmospheric Pressure MALDI

The proteomics approach targeting species-unique peptides derived from most abundant proteins (Warscheid and Fenselau 2004) for bacteria identification was recently improved by using an ion trap MS with an atmospheric pressure MALDI source to increase the speed of analysis and accurate precursor ion selection (Pribil et al. 2005).

There are several advantages afforded by AP-MALDI over vacuum MALDI. First, the fact that AP-MALDI operates at atmospheric pressure offers the opportunity to utilize liquid matrices and matrices that are not vacuum compatible. This makes AP-MALDI a very versatile instrument for a variety of samples. Second, the AP-MALDI method has shown indications to be an even "softer" ionization technique than vacuum MALDI due to fast thermalization of the ion internal energy at atmospheric conditions. This feature is important in the analysis of labile biomolecules and noncovalent complexes. Third, AP-MALDI is decoupled from the mass analyzer. This decoupling allows AP-MALDI to be interfaced with ion trap instruments or any other MS system equipped with an API (typically those which use ESI). Ion trap instruments allow MS/MS analysis, which is much more powerful and easier to perform than PSD using TOF instruments. Finally, AP-MALDI is the most

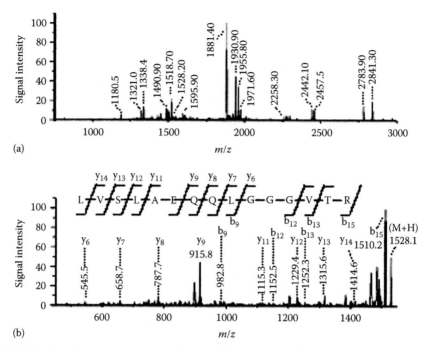

FIGURE 4.1 Partial mass spectra from a 1:1 mixture of *B. anthracis* str. Sterne and *B. subtilis* str. 168 treated with acid and digested with immobilized trypsin in situ. (a) Survey scan of the mixture. SASP peptides specific for *B. anthracis* and *B. subtilis* are indicated by closed and open diamonds, respectively. Peptides specific for the *B. cereus* group are indicated by the symbol #. (b) Tandem mass spectrum and interpretation of the *B. anthracis*–specific peptide with *m/z* 1528.1.

cost-effective way to get MS/MS data from MALDI ions. The examples of MALDI spectra obtained during analysis of bacterial mixture composed of *B. anthracis* and *B. subtilis* cells, and targeting SASPs are shown in Figure 4.1.

4.2.3 LC-ESI-MS/MS

In general, the complexity of peptide mixtures created during shotgun digestion of proteins extracted from bacterial cells dictates the use of strategies intended to diminish the complexity of peptide ions introduced into a MS (VerBerkmoes et al. 2004). The most popular of these approaches, called multidimensional protein identification technology, uses a column containing two different separation materials and multiple steps of chromatography for separation of peptides (Wolters et al. 2001). However, this method is not optimal for peptide separation due to the elution of peptides with a solvent step gradient during ion-exchange chromatography. Therefore, off-line techniques based on a continuous gradient ion-exchange separation of peptides that are subsequently analyzed by reversed-phase LC (RPLC) coupled with ESI-MS/MS represent a better choice for the comprehensive analysis of the bacterial proteome. Using this approach, Jaffe et al. (2004) found almost 10,000 unique

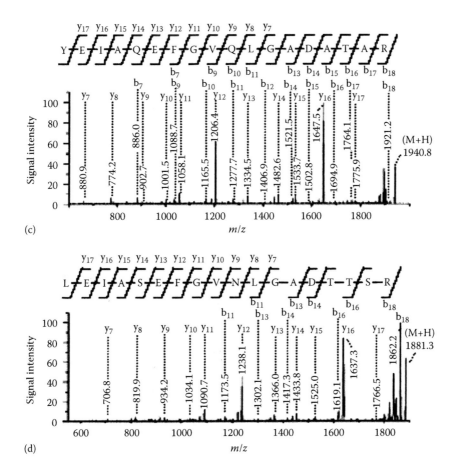

(c)

(d)

FIGURE 4.1 (continued) Partial mass spectra from a 1:1 mixture of *B. anthracis* str. Sterne and *B. subtilis* str. 168 treated with acid and digested with immobilized trypsin in situ. (c) Tandem mass spectrum and interpretation of the *B. cereus* group-specific peptide with *m/z* 1940.6. (d) Tandem mass spectrum and interpretation of the *B. subtilis*–specific peptide with *m/z* 1881.3 (Pribil et al. 2005).

tryptic peptides corresponding to 81% of the predicted open reading frames (ORFs) for *Mycoplasma pneumoniae*.

The sequencing of only a few peptides derived from a given ORF of a known organism is usually sufficient for the protein identification. Therefore, methods based on the targeted fractionation of peptides, such as the presence of a particular amino acid or chemical group (e.g., sulfhydryls of cysteinyl residues), are also capable of substantially reducing peptide complexity in a mixture. However, the purpose of these fractionation methods is to increase the probability that peptides from a broader range of proteins, including those from low copy number proteins, will be selected for sequencing. Hence, the simplest way to achieve this goal is the application of a gas-phase fractionation (GPF) procedure. GPF takes advantage of a MS capability to select ions for CID-based sequencing in a narrow mass range. Focusing on a narrow mass range during multiple injections of sample aliquots allows a greater number of peptide ions to be analyzed and substantially increases the coverage of an investigated bacterial proteome (Jabbour et al. 2005; Kolker et al. 2003, 2005).

Thus, a gel-free proteomic procedure based on the LC-ESI-MS/MS of peptides generated from cellular proteins is an attractive platform for large-scale analyses of bacterial proteomes (Corbin et al. 2003; Taoka et al. 2004) and can be utilized for the identification and classification of microorganisms (Dworzanski et al. 2004a,b, 2005, 2006; Dickinson et al. 2005).

Recently, a microfluidic chip for peptide analysis has been introduced on the market by Agilent Laboratories in Palo Alto, CA (Yin et al. 2005), that simplifies the nano-LC/MS system. In this approach, the traditional use of an enrichment column, a separation column, a nanospray tip, and the fittings needed to connect these parts together has been replaced by a microfabricated device. A single microfluidic chip integrates these components, thus eliminating the need for conventional LC connections. The chip was fabricated by laminating polyimide films with laser-ablated channels, ports, and frit structures (Figure 4.2). The enrichment and separation columns were packed using conventional reversed-phase chromatography particles. A face-seal rotary valve provided a means for switching between sample loading and separation configurations with minimum dead and delay volumes while allowing high-pressure operation (Figure 4.3). The LC chip and valve assembly were mounted within a custom electrospray source on an ion trap MS. The overall system performance was demonstrated through reversed-phase gradient separations of tryptic protein digests at flow rates between 100 and 400 nL/min. Microfluidic integration of the nano-LC components enabled separations with subfemtomole detection sensitivity, minimal carryover, and robust and stable electrospray throughout the LC solvent gradient.

4.2.4   CE-ESI-MS/MS

ESI-MS/MS allows for online detection and identification of peptides separated by CE (Janini et al. 2003a,b); however, this separation mode was rarely used for bacteria identification purposes. Recently, a Taiwanese group described a successful use of this technique for identification of microbial mixtures using quadrupole ion trap operated in a selective MS/MS mode.

Hu et al. (2005, 2006) in two published reports used selective MS/MS analysis targeting species-unique peptide ions from selected bacteria, derived from tryptic peptides which were separated by CE-MS/MS. For that purpose, they first built a small database of proteotypic tryptic peptides that are (a) species-specific biomarkers for targeted strains and (b) derived from abundant proteins. Isolated ions of such peptides were analyzed by using a selective reaction monitoring (SRM) approach. In this approach, the wideband excitation energy was set to 42% of the normalized collision energy and lasted for 30 ms. Mass range of monitored product ions was set to the allowable maximum value to cover the dissociation fragments.

(a)

(b)

FIGURE 4.2 Microfluidic chip for peptide analysis with an integrated HPLC column, sample enrichment column, and nano-electrospray tip. The chip clamping mechanism: (a) schematic of chip with clamp in open position consisting of the stator, clamp, and valve assembly and (b) photo of chip in the clamped position ready for rotation into the spray. (From Yin, H. et al., *Anal. Chem.*, 77(2), 527, 2005.)

In this approach, several different ions are scanned sequentially and the process is repeated throughout the whole data acquisition time. The example of selected peptides used for detection of targeted bacteria is shown in Table 4.1.

The overall identification success for this method was found to be 97% based on the analysis of 34 clinical samples. However, these clinical specimens were subjected to CE/MS analysis after direct cultivation aimed to increase the bacterial mass. The total analysis time was ca. 8 h and included a 6 h long cultivation step, while the time-consuming digestion process was shortened to 15 min by the application of microwave-assisted proteolysis (Lin et al. 2005a,b).

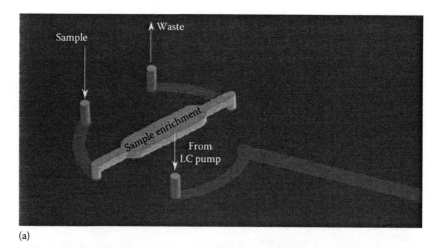

(a)

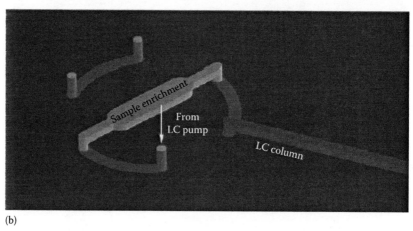

(b)

FIGURE 4.3   Schematic of a chip in the loading process: (a) the sample loading configuration of the rotor channels and (b) the LC running configuration of the rotor channels. (From Yin, H. et al., *Anal. Chem.* 77(2), 527, 2005.)

4.2.5   PROTEIN DATABASES

MS-based methods for protein identification depend on the availability of a protein database for the investigated organism, because both peptide masses from MALDI-MS experiments and uninterpreted product ion mass spectra of peptide ions should be matched to database sequences of predicted proteins (Aebersold 2003). Although the de novo interpretation of product ion mass spectra of peptides can reveal amino acid sequences (Standing 2003), they are also used as a query by database search programs based on sequence similarity (e.g., FASTA, BLAST, MS BLAST) (Shevchenko et al. 2002; Wheeler et al. 2003; Habermann et al. 2004; Zhong and Li 2005). However, this approach also allows for cross-species protein identification if sequences from homologous proteins of closely related organisms are included in a database (Liska and Shevchenko 2003).

TABLE 4.1
Selected Peptides Obtained from Data-Dependent Experiments of Each Pathogen

Bacterium	Isolated Ion	Sequence	Protein	MW	Peptide Position
S. aureus	1076.52 (1+)	-ETVGNVTDNK-	Hypothetical protein	6,996	16–25
	1262.60 (1+)	-SGEESEVLVADK-	Hypothetical protein	21,243	92–103
Staphylococcus epidermidis	1040.49 (2+)	-YGPVDGDPITSTEEIPFDK-	Accumulation-associated protein	157,025	1086–1104
	949.99 (2+)	-AHLVDLAQHNPEELNAK-	Immunodominant antigen A	24,514	37–53
P. aeruginosa	1638.79 (1+)	-IEDTDFAAETANLTK-	Flagellin	34,341	307–321
	1036.04 (2+)	-TVIHTDNAPAAIGTYSQAIK-	Conserved hypothetical protein	13,583	4–23
S. pyogenes	1087.56 (1+)	-EAVEGAVDAVK-	Hypothetical protein SPy2005	6,939	52–62
	1374.69 (1+)	-HGELLSEYDALK-	M protein	11,914	54–65
S. typhimurium	1968.99 (1+)	-AQPDLAEAAATTTENPLQK-	Phase 1 flagellin	41,275	340–358
	1272.67 (2+)	-TLHLADSELSEEALIQALVEHPK-	Putative arsenate reductase	13,351	70–92
S. pneumoniae	1451.74 (1+)	-DFHVVAETGIHAR-	Phosphocarrier protein of the PTS	8,916	5–17
	1521.80 (1+)	-TVGDLVAYVEEQAK-	Acyl carrier protein	8,245	61–74
E. coli	1833.97 (1+)	-EAAIQVSNVAIFNATTGK-	50S ribosomal protein L24	11,321	62–79
	1681.78 (1+)	-EAIGYADSVHDYVSR-	Bacterioferritin	18,495	103–117
Enterococcus faecalis	955.93 (2+)	-TLEEGQAVTFDVEDSDR-	Cold-shock domain family protein	4,949	29–45
	1935.93 (1+)	-TLEEGQAVTFEIEEGQR-	Cold-shock domain family protein	7,163	40–56
Streptococcus agalactiae	1308.67 (1+)	-VTVEVTYPDGTK-	Alpha-like protein 4	38,597	286–297
	1073.55 (1+)	-DAVEGAVDAVK-	Hypothetical protein	6,999	52–62

The availability of almost 600 fully sequenced microbial genomes (as of November 2007) together with more than 700 prokaryotic genome-sequencing projects in progress provides an unprecedented resource for proteomics studies because protein databases are derived from genomic sequences. In fact, amino acid sequences in these databases represent a conceptual translation of nucleotide sequences in computationally determined ORFs that potentially encode proteins. For clarity, ORF should be understood as a computationally predicted section of a DNA sequence that begins with an initiation codon and ends with a stop codon. Therefore, each ORF has the potential to encode a single polypeptide that may be expressed as a protein; however, many may not actually do so. Furthermore, a protein should be understood as one of many isoforms representing the expressed and matured gene product that may be substantially different from a polypeptide specified by a nucleotide sequence. Generally, these differences are not rare and originate mainly from co-translational modifications or PTM of a nascent polypeptide. Co-translational modification refers to the removal of N-terminal methionine by N-methionylaminopeptidase and affects the majority of bacterial proteins. PTMs comprise both the proteolytic processing of a polypeptide and covalent modifications of its amino acids (Hesketh et al. 2002). Therefore, the available database searching algorithms identify ORF as a protein that is not mature. Moreover, during analysis of an unknown bacterium, the confirmation of the full amino acid sequence or "100% coverage" of a potential protein would be required for the identification of an ORF. Although this requirement may be relaxed under certain conditions such as the proteomic analysis of a known type strain with a fully sequenced genome, the true identification of proteins is rarely achievable during high-throughput analyses of bacterial proteomes.

There are many reasons why this occurs, and they may be divided into two categories. The first includes factors associated with sample preparation, peptide ionization, and MS fragmentation processes, while the second category includes biological factors such as PTMs. For example, most exported proteins are synthesized as precursors with an N-terminal signal peptide that is removed during the translocation process. Although a signal peptide sequence is present in a database "protein," the theoretically expected tryptic peptides of such proteins will not be produced from analyses of expressed proteins. Although this and other common proteolytic modifications do not change amino acid sequences in polypeptide products, there are numerous exceptions to this rule. For example, some bacterial proteins contain internal segments of amino acids (called inteins) that self-catalyze their excision from the host protein and ligate the flanking fragments by a peptide bond. In this process that is analogous to the excision of introns on the mRNA level, two new proteins are formed: a mature host protein and the free intein (Amati et al. 2003). Furthermore, although introns were assumed absent in genes coding bacterial proteins, such intervening sequences were reported recently in bacterial genes from *Clostridium difficile* (Braun et al. 2000), *B. anthracis* (Ko et al. 2002), and *Actinobacillus actinomycetemcomitans* (Tan et al. 2005).

Although PTMs of amino acids do not change their sequence, they "decorate" proteins by specific covalent attachments that add extra mass. The most frequently occurring PTM appears to be phosphorylation, which is an important regulation mechanism controlled by protein-specific phosphorylating and dephosphorylating enzymes (kinases and phosphatases, respectively). However, many other PTMs have been widely documented, including N-acetylation, methylation, thiomethylation, adenylation, glycosylation, and myristoylation, which are common mechanisms for modulating structural and functional properties of bacterial proteins (Benz and Schmidt 2002). In summary, it is not surprising that only in rare cases molecular weights and full amino acid sequences of proteins predicted from the annotated ORF are really observed at the level of expressed proteins (Wang et al. 2002a,b).

4.2.6   CLASSIFICATION AND IDENTIFICATION OF BACTERIA BASED ON THE DISTRIBUTION OF PHYLOGENETIC PROFILE

With currently available commercial MS, ca. 6–30 amino acid long peptide segments of proteins can be fragmented and mass analyzed in an ~200 ms period. Database searches with the acquired product ion mass spectra provide amino acid sequences of peptides that represent genomic information translated from matching ORF segments in all database bacteria. However, it is still a challenging task to translate the raw data generated from high-throughput MS experiments into biologically meaningful and easy-to-interpret results.

Although the identification of proteins is helpful in establishing the identity of an analyzed bacterium, the sequence coverage of so-called "identified proteins" is rarely complete during shotgun sequencing. This is caused by factors related to sample preparation, the mass spectral acquisition mode, and biological reasons (vide supra). Moreover, even full sequence coverage of a protein merely indicates that sequence is the same for only one of many of gene products. Thus, a more reliable method to determine the overall genomic similarities between a test sample and database bacteria would be to infer them from a set of confidently identified peptide sequences mapped to diverse chromosomal locations.

To achieve such a goal, peptide sequences were identified from product ion mass spectra during analysis of an unknown sample and then assigned to database bacteria (Dworzanski et al. 2004a,b, 2006). Histograms were constructed for the peptides matching each bacterium and were used to reveal the closest bacterial database relatives. The highest number of confidently identified peptides and comparative analysis of peptide-to-bacterium assignments was used to identify the test sample. To simplify the identification process, other researchers (VerBerkmoes et al. 2005) used only those sequences that were uniquely identified only in one database species. In many cases, these data processing methods allowed for unequivocal identification of investigated bacteria; however, both approaches are not generally applicable.

The integration of MS-based proteomic techniques with the bacterial taxonomy database and numerical taxonomy methods can provide a more complete, deeper understanding of relationships among bacteria. These methods extend the raw information provided by simple searching protein databases with uninterpreted product ion mass spectra (Dickinson et al. 2005; Dworzanski et al. 2005; Jabbour et al. 2010a,b; Deshpande et al. 2011).

The general strategy used for the identification of bacteria is schematically repre-sented in Figure 4.4 (Dworznaski et al. 2005). First, the preparation of tryptic peptides is performed using well-established protocols that rely on cell lysis (e.g., sonication) followed by denaturation of proteins and the optional reduction of the cysteine disul-fide bonds in the protein chains by carboxyamidomethylation. Peptides are obtained by trypsin proteolysis with subsequent LC-ESI-MS/MS analysis. During standard 1D- or 2D-LC-ESI-MS/MS analyses, separated peptides are electrosprayed into a MS. The most abundant ions in the precursor scan are automatically selected for fragmentation via CID. The recorded tandem mass spectra are processed to identify the amino acid sequences of the precursor peptide ions.

There are many commercially available algorithms including SEQUEST for database searching using tandem mass spectra (Sadygov et al. 2004). The SEQUEST algorithm that matches uninterpreted product ion mass spectra recorded during pep-tide analysis with theoretical fragmentation patterns predicted for all tryptic peptide sequences in a protein database was chosen to determine the best fit. Therefore, the search results depend significantly on the composition and quality of a database. Different laboratories used protein databases comprised of diverse subsets of pro-tein sequences available in public databases (Harris and Reilly 2002; Dworzanski et al. 2004a,b; Warscheid and Fenselau 2004; VerBerkmoes et al. 2005). However, a curated database comprising only sequences from ORFs annotated during complete sequencing projects of bacterial genomes seems to be the most appropriate to infer identities and for comparative analyses of an unknown bacterium in a systematic manner. Hence, the database used to illustrate the present strategy (Figure 4.4) was constructed from computationally predicted proteomes of all bacteria with fully sequenced genomes that are available from the National Institutes of Health National Center for Biotechnology Information (NCBI) Internet ftp site ftp://ftp.ncbi.nih.gov/genomes/Bacteria.

To speed up searches of database sequences, proteins are usually digested in silico following the cleavage rules of the protease applied for sample processing. Tryptic peptide sequences stored in the database can be viewed as a virtual array of peptide "probes" composed of tens of millions of elements that are interrogated by SEQUEST to determine matches between experimental peptide sequences derived from the investigated proteome and the database peptides. Currently, 548 eubacte-rial and 47 archaeal genomes and many of their plasmids are fully sequenced and available (as of October 31, 2007) on the Internet. For instance, fully assembled genomes of Eubacteria represent strains from 85 genera, 61 families, 41 orders, 20 classes, and 12 phyla classified in accordance with the accepted taxonomy for each strain. Therefore, each database protein sequence may be supplemented with taxo-nomic information on a source microorganism and the chromosomal position of each respective ORF.

A database search algorithm like SEQUEST attempts to match every experimen-tal spectrum to theoretical spectra of database peptides. There is a need to use well-defined criteria to determine the validity of each automated assignment. The simplest way to express the accuracy of such assignments would be to calculate the probabil-ity that a given match is correct. Although there are many computational ways to determine such probabilities (Sadygov et al. 2004), the PeptideProphet algorithm

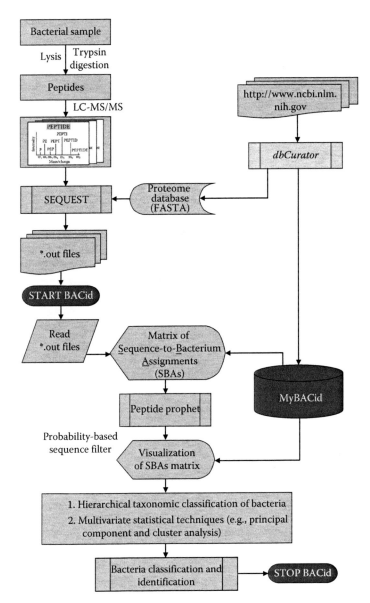

FIGURE 4.4 A simplified flowchart summarizing processing procedures of product ion mass spectra from an LC-ESI-MS/MS analysis of tryptic bacterial peptides. (From Dworzanski, J.P. et al., *Proceedings of the 53rd ASMS Conference on Mass Spectrometry and Allied Topics*, San Antonio, TX, TP22, 2005.) The SEQUEST algorithm (ThermoFinnigan) used the product ion mass spectra to search a curated bacterial proteome database assembled by a module "*dbCurator*" from publicly available sources. The search outputs (*out files) were read by the BACid in-house software that arranges a STB assignments matrix using taxonomy information stored in the MyBACid local database. The assignments were filtered based on the probability of correct spectrum-to-sequence matches computed by a program "PeptideProphet™." (From Keller, A. et al., *Anal. Chem.*, 74(20), 5383, 2002.)

developed at the Institute of Systems Biology (Keller et al. 2002) was used by the ECBC Point Detection Team in the presented application. This program determines the probability of a correct peptide assignment to a database sequence; therefore, only peptides identified with high confidence are selected for further comparative analyses. These peptides represent the peptide profile of a tested microorganism.

However, each peptide sequence may be found in one or more proteome/genome of database bacteria, and such assignments form a phylogenetic profile of a peptide. These profiles form a matrix of peptide-to-bacterium assignments that can be visualized as a bitmap and analyzed to determine relationships between a test sample and database microorganisms. This step may be performed using a set of common statistical techniques for analysis of multivariate data.

4.2.7 CLASSIFICATION OF A BACTERIAL TEST SAMPLE

To illustrate this process, a bacterial strain isolated from a rice dish and identified as *B. cereus* serotype H10 was chosen as a test sample and analyzed using the LC-ESI-MS/MS technology (Dworzanski et al. 2007). This strain was previously characterized (La Duc et al. 2004) and showed a high nucleotide sequence similarity of its 16S rDNA gene to *B. cereus* and *B. anthracis* type strains of 99.3% and 99.7%, respectively. However, a substantially higher percentage of the DNA–DNA hybridization was observed with *B. cereus* (72%) than with *B. anthracis* (50%). In addition, the sequencing of a *gyrB* gene indicated 99.1% and 90.9% similarity to *B. cereus* and *B. anthracis*, respectively, and the experimental *B. cereus* strain displayed a lack of genes encoding the protective antigen (*pag*) and capsular antigen (*cap*) that are virulence determinants for *B. anthracis* (La Duc et al. 2004).

Uninterpreted product ion mass spectra recorded for the *B. cereus* serotype H10 sample were searched against a protein database (Figure 4.4). The SEQUEST output files were processed to yield a set of 203 peptides that with high probability (99%) were correctly identified. A matrix of STB assignments composed of phylogenetic profiles of the 203 peptides was used for the identification of the test sample. A bioinformatics procedure used the established taxonomic classification scheme of bacteria (Figure 4.5) and analyzed the similarity of the test sample to database strains grouped in accordance to their taxonomic position (phyla, classes, orders, families, genera, species). Because bacteria represented in a database belong to 12 phyla in the first step a similarity coefficient was computed that represented a fraction of the accepted peptides that match bacteria in each phylum. A histogram of the similarity coefficients (Figure 4.5a) indicates that the bacterium in the test sample belongs to *Firmicutes*. By repeating this procedure, the class of the sample was identified as *Bacilli*, the order was *Bacillales*, and the family was identified as *Bacillaceae*. However, the *Bacillaceae* family is represented in the database by only 13 strains. Twelve belong to the genus *Bacillus*, and one strain represents the genus *Oceanobacillus*. Similarity coefficients between the test sample and database *Bacillaceae* strains are shown in Figure 4.5b. This bottom histogram in Figure 4.5 shows that the test sample has the highest similarity to *B. cereus* ATCC 14579 (species 1). The values of the similarity coefficient between the test sample and species 1–8 in Figure 4.5b were in the range of 65%–92% while the remaining

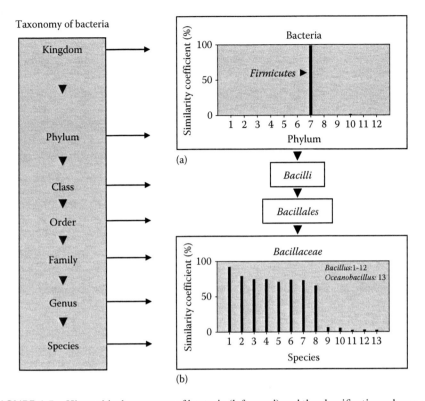

FIGURE 4.5 Hierarchical taxonomy of bacteria (left panel) and the classification scheme of a *B. cereus* Serotype H10 isolate test sample (right panel). Sequences of 203 tryptic peptides identified during LC-MS/MS analysis of *B. cereus* ser. H10 were assigned to 202 database bacteria. Similarities between the test sample and database strains, grouped in accordance with their taxonomic position in the classification hierarchy, were estimated by comparing similarity coefficients. Similarity coefficients were computed as a fraction of peptides that matched bacteria in a given taxon (number of positive matches × 100/total number of identified peptides). (a) Histogram of similarity coefficients obtained for database bacteria grouped in 12 phyla places the test sample in *Firmicutes*. [Phyla: 1—*Actinobacteria*, 2—*Aquificae*, 3—*Bacteroidetes/Chlorobi* group, 4—*Chlamydiae*, 5—*Cyanobacteria*, 6—*Deinococcus/ Thermus*, 7—*Firmicutes* (F), 8—*Fusobacteria*, 9—*Planctomycetes*, 10—*Proteobacteria*, 11—*Spirochaetes*, and 12—*Thermatogae*]. Further analysis of assignments to *Firmicutes* identified that the class of the test sample was *Bacilli*, the order was *Bacillales*, and the family was identified as *Bacillaceae*. (b) The *Bacillaceae* family is represented in the database by 12 strains from the genus *Bacillus* (1–12) and one strain from *Oceanobacillus* (13). Similarity coefficients between the test sample and the following database *Bacillaceae* strains were computed: (1) *B. cereus* ATCC 14579; (2) *B. cereus* ATCC 10987; (3) *B. anthracis* Ames; (4) *B. anthracis* Ames ancestor; (5) *B. anthracis* Sterne; (6) *B. thuringiensis* 97–27; (7) *B. cereus* ZK; (8) *B. anthracis* A 2012; (9) *B. licheniformis*; (10) *B. subtilis*; (11) *B. clausii*; (12) *B. halodurans*; (13) *Oceanobacillus iheyensis*. Histogram of similarity coefficients indicates that strains 1–8 and 9–13 form two separate clusters, and *B. cereus* ATCC 14579 (strain #1) shows the highest similarity to the test sample. Peptide assignments to all *Bacillaceae* strains were further analyzed using hierarchical cluster analysis (see Figure 4.6).

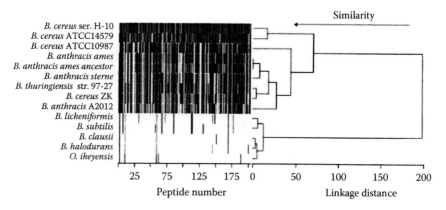

FIGURE 4.6   "Gel view" of the binary peptide-to-bacterium assignment matrix (left panel) and a dendrogram obtained after applying hierarchical cluster analysis (right panel) to the matrix of 203 peptide sequences assigned to the test sample and 13 database strains of the *Bacillaceae* family. The test sample represents an isolate of *B. cereus* serotype H10 strain, and the error rate of peptide identification for this analysis was lower than 1%. A furthest neighbor-joining algorithm was applied, and the squared Euclidean distance was chosen for determining the inter-cluster distances. This linkage distance is equivalent to the number of peptides differentiating bacterial strains or their clusters. The visualized relationships between the test sample and *Bacillaceae* strains indicate that the test sample clusters with the *B. cereus* group of bacteria and forms the sub-cluster with a type strain *B. cereus* ATCC 14579. However, the distance between these two strains, measured as 16 peptide sequences that represent 8% of all identified peptides, indicates that the difference is statistically significant. Therefore, the test sample can be identified as the *B. cereus* strain with a 92% similarity to the closest relative in the database, i.e., strain ATCC 14579.

Bacillaceae strains showed similarities lower than 6.6%. However, 8% of peptides differ between the most similar database strain (species 1; *B. cereus* ATCC 14579) and the test sample, and the difference is statistically significant (the error rate of sequence assignments for the dataset was lower than 1%). Therefore, results of the *B. cereus* analysis are in full agreement with its taxonomic position as determined by the established genomic methods.

Figure 4.6 provides a visualization of the peptide-to-bacterium assignment matrices in a "gel-view" format for the *Bacillaceae* family of species shown in Figure 4.5b and is accompanied by results of a hierarchical clustering analysis shown as a dendrogram. The dendrogram in Figure 4.6 clearly indicates that the test sample (*B. cereus* serotype H10) clusters with the taxonomic *B. cereus* group of database species that comprises *B. cereus*, *B. anthracis*, and *B. thuringiensis* strains. Moreover, due to the high discriminative power, this proteogenomic technique can also be used for typing and identification of bacteria at the strain level because the test strain is clearly distinguished from other *B. cereus* strains (Figure 4.6).

Because the differences in Figure 4.6 suggest specific testable predictions regarding phenotypic differences between the test sample and its closest relatives, this type of investigation can easily be augmented to reveal target candidates for the development of potential vaccines, diagnostics, or therapeutics.

4.3   NUCLEIC ACIDS

MS identification of bacteria based on genomic information may be revealed through sequencing amino acids derived from microbial proteins that take advantage of the availability of a few hundred fully sequenced bacterial genomes or through direct analysis of nucleic acids. In the latter case, additional sources of genomic information, which comprise nucleotide sequences of ribosomal ribonucleic acids (rRNA) or noncoding regions of DNA, are usually used in a manner that takes advantage of the PCR amplification options available for nucleic acid–based approaches. For example, there are well-established methods to infer phylogenetic relationships among bacteria determined by 16S rRNA alignments and large numbers of these sequences (ca. 180,000) are freely available. Amplification of species-specific sequences, the use of nucleotide microarrays, and in situ hybridization are based on the presence of unique subsequences in the target sequence and, therefore, require prior knowledge of what organisms are likely to be present in a sample. Furthermore, MS is not limited by a pre-synthesized inventory of probe/primer sequences.

MALDI-TOF-MS has been used for chain-termination sequencing of nucleic acids; however, the maximum read length using such an approach is ca. 60 nucleotides (Kwon et al. 2002). In general, very high-resolution measurements are required for unambiguous compositional assignments (±1 ppm) to calculate base composition for PCR products using ESI-Fourier transform ion-cyclotron resonance (ESI-FTICR) as has been demonstrated by Hofstadler et al. (2005) Unfortunately, the resolution required for unambiguous compositional assignment (±1 ppm) of such large molecules requires instrumentation that is out of reach for many laboratories. Thus, it is advantageous to introduce a cleavage step, which reduces the resolution requirements while retaining valuable information.

4.3.1   RNA

Analysis of single-stranded nucleic acids is generally preferred as the same information content is available at roughly half the mass. Furthermore, endoribonucleases can be used to selectively cleave a single-stranded RNA after a particular base (for example, guanosine residues in the case of RNase T1). Despite the information loss associated with compositional rather than sequential analysis, microbial identification based upon the composition of base-specific cleavage products appears extremely promising. For example, Von Wintzingerode et al. (2002) described the comparison of base-specific cleavage patterns derived from *Bordetella* species against the patterns predicted by virtual cleavage of 50 published 16S rDNA sequences, including 13 sequences which were known to be closely related. Discriminating masses were compared and strains were typed by inspection (Figure 4.7). In addition, the same group of researchers used similar methods to rank the identification of mycobacteria (Lefmann et al. 2004).

This relatively rapid approach to the 16S rRNA gene (16S rDNA)-based bacterial identification combines uracil-DNA glycosylase (UDG)–mediated base-specific fragmentation of PCR products with MALDI-TOF-MS. 16S rDNA signature sequences were PCR-amplified from both cultured and as-yet-uncultured bacteria in the presence

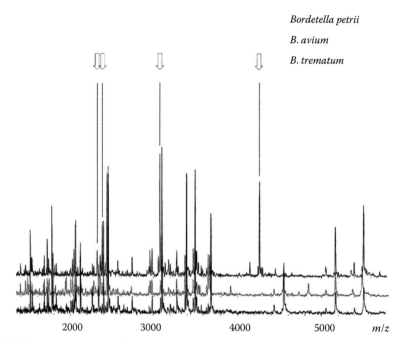

FIGURE 4.7   Overlay of mass spectra of *Bordetella petrii*, *Bordetella avium*, and *Bordetella trematum*. Peaks specific for *B. petrii* are marked by the arrows. For display, mass spectra were smoothed by a five-point average (Savitzky–Golay method) and baseline-subtracted. (From Von Wintzingerode, F. et al., *Proc. Natl. Acad. Sci. U.S.A.*, 99(10), 7039, 2002.)

of dUTP instead of dTTP. These PCR products were then immobilized onto a streptavidin-coated solid support to selectively generate either sense or antisense templates. Single-stranded amplicons were subsequently treated with UDG to generate T-specific abasic sites and fragmented by alkaline treatment. The resulting fragment patterns were analyzed by MALDI-TOF-MS. Mass signals of 16S rDNA fragments were compared with patterns calculated from published 16S rDNA sequences. MS of base-specific fragments of amplified 16S rDNA allows reliable discrimination of sequences differing by only one nucleotide. This approach is fast and has the potential for high-throughput identification as required in environmental microbiology (Von Wintzingerode et al. 2002).

This technique uses well-established genotypic markers, the 16S rDNA, allowing universal identification of both cultured and as-yet-uncultured bacteria (Ludwig et al. 1998). PCR-amplified signals by MS analysis of 16S rDNA-specific fragmentation patterns can be superior to protein- or whole cell–based identification by MS that may require in vitro cultivation of bacterial cells as a method aimed to increase cellular mass before analysis.

MALDI-TOF-MS analysis of base-specific 16S rDNA fragmentation patterns (as well as sequencing of proteins and most other DNA-based identification methods) is not biased by the cultivation conditions. The absence of any time-consuming electrophoresis or chromatography step is an important advantage of MS analysis when compared with conventional 16S rDNA sequencing or rDNA fingerprinting by restriction analysis. Other sequencing techniques such as pyrosequencing have been proposed recently as viable alternatives to gel- or capillary-DNA sequencing. However, their short read lengths of 20–30 nucleotides strongly limit the use for 16S rDNA-based bacterial identification, because additional typing steps are required, e.g., the use of taxon-specific PCRs. Compared with these techniques, the principle of base-specific fragmentation of PCR amplicons combines the possibility of scanning larger sequence regions for bacterial typing with the speed and accuracy of MALDI-TOF-MS.

The experimental procedure required for PCR amplification and subsequent base-specific fragmentation is rather simple and could make this method an ideal candidate for high-throughput identification in clinical diagnostics and environmental microbiology. It has great potential for bacterial and fungal pathogens that cause problems in routine conventional diagnostics. The MS-based identification approach is not restricted to 16S rDNA but can be expanded to other genotypic markers, e.g., typing of viral strains or typing of other specific regions, which will broaden its applicability even further.

Jackson et al. (2007) described successful organism identification by MS of 16S ribosomal RNA cleavage products which relies on comparison of observed masses to those predicted for all previously sequenced taxa. The authors took advantage of the public availability of large numbers of bacterial 16S rRNA sequences that in the past has facilitated microbial identification and classification using nucleic acid hybridization and other molecular approaches. In short, their MS analysis of base-specific cleavage products uses an efficient coincidence function for rapid spectral scoring against a large database of predicted masses of 16S RNA products. Using this approach in conjunction with universal PCR of the 16S rDNA gene, four bacterial isolates and an uncultured clone were successfully identified against a database of predicted cleavage products derived from over 47,000 16S rRNA sequences representing all major bacterial taxa. At present, the conventional DNA isolation and PCR steps require ~2 h, while subsequent transcription, enzymatic cleavage, MS analysis, and database comparison require <45 min. All steps are amenable to high-throughput implementation.

Figure 4.8 displays comparison of spectra acquired from digests of amino–allyl U-modified RNA from the "AB" sequence regions of two different organisms, *P. aeruginosa* and *Vibrio proteolyticus*. The spectra are shown as raw data with only the single point mass calibration performed and the resolution at full-width half-maximum of the major peaks at the 500–800 range is observed that is typical for operation in linear mode. At this resolution, without amino–allyl U modification, many products having only a U/C difference in composition would be superimposed. Instead, U/C "compomers" differ by ca. 55 Da due to the amino–allyl group located on the 5-position of the uridine base. For organism identification, spectra are processed beyond the raw data depicted in Figure 4.8.

Table 4.2 gives the resulting organism identification, using the circumscribed approach, compared to those obtained by BLAST against the entire NCBI nucleotide–nucleotide database of sequences determined by conventional CE

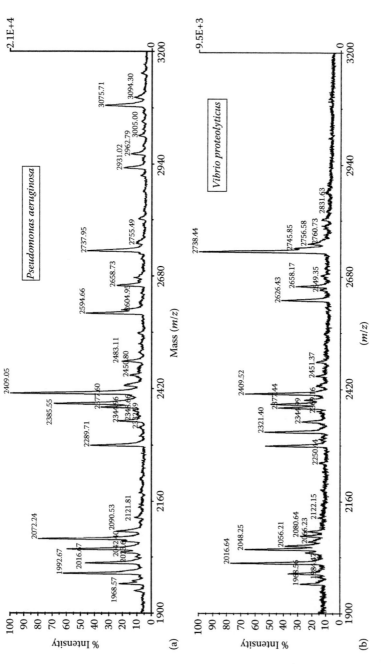

FIGURE 4.8 (a) and (b) Comparison of typical spectra of base-specific cleavage patterns acquired from two different organisms. Only single point calibration has been performed on the raw spectra presented. The masses 2377 and 2385 Da corresponding to the mass-modified RNA oligonucleotides, UAAUACG, and AUCCUUG, respectively, are illustrative of nearest mass neighbors under conditions of 100% amino–allyl U substitution. (From Jackson, G.W. et al., *Int. J. Mass Spectrometry*, 261, 218, 2007.)

TABLE 4.2
Comparison of Bacterial Identification by Conventional Sequencing vs.
Base-Specific MS Coincidence Analysis

Sequencing/BLAST Score, "Bits" (Rank)	MS Coincidence Analysis Score (Rank)
Sample: *E. coli* (K-12, MG1655)	
E. coli K-12 MG1655 930 (1)	*E. coli*; K-12; M87049 0.476 (1)
E. coli strain RW-29 16S ... 930 (1)	*E. coli*; K-12; U18997 0.476 (1)
E. coli O157:H7 EDL933 930 (1)	*E. coli*; O157:H7; BA000007 0.476 (1)
E. coli 16S rRNA gene 930 (1)	*E. coli*; AU1713; AY043392 0.476 (1)
E. coli C2 16S rRNA 930 (1)	*E. coli*; L10328 0.476 (1)
E. coli 16S rRNA gene 930 (1)	*E. coli*; CCCO4; AF511430 0.476 (1)
Escherichia albertii strain 10457 ... 930 (1)	*Shigella flexneri* 2a str. 301; AE005674 0.476 (1)
E. albertii strain 12502 ... 930 (1)	*S. flexneri* 2a str. 2457T; AE016989 0.476 (1)
Sample: *P. aeruginosa* (ATCC 25102)	
P. aeruginosa gene for 16S rRNA 946 (1)	*P. aeruginosa*; AT10; AJ549293 0.471 (1)
Pseudomonas sp. pDL01 16S rRNA 938 (2)	*Pseudomonas alcaligenes* (T); LMG 1224T 0.455 (2)
Pseudomonas sp. Bxl-1 938 (2)	*P. alcaligenes*; M4-7; AY835998 0.446 (3)
P. aeruginosa ATCC BAA-1006 938 (2)	*P. aeruginosa*; ATCC BAA-1006; 0.439 (4)
Pseudomonas sp. BWDY-42 16S rRNA 938 (2)	*P. aeruginosa*; PAO1; AE004949 0.439 (4)
P. aeruginosa partial 16S rRNA 938 (2)	*P. aeruginosa*; ATCC 27853; 0.439 (4)
Pseudomonas sp. HY-7 16S rRNA 938 (2)	*P. aeruginosa*; SCD-13; 0.439 (4)
Pseudomonas sp. LQG-3 16S 938 (2)	*Pseudomonas* sp. pDL01; AF125317 0.439 (4)

Source: Jackson, G.W. et al., *Int. J. Mass Spectrometry*, 261, 218, 2007.

sequencing of the same amplicons. All BLAST scores (bits) were taken as reported by the web site without further modification. As can be seen, MS was quite successful in identifying these bacteria when compared to full sequence analysis. Note that only a single cleavage reaction and only the sense strand of each amplicon were used to obtain this result. In the case of *E. coli* correct identification to the strain level was achieved, while *P. aeruginosa* was correctly identified to the species level (the strain used by investigators was not represented in databases). However, sequencing results indicated only 11 nt differences over the combined 918 nt sequence region between the analyzed strain (ATCC 25102) and the top strain identified by MS (Jackson et al. 2007).

These examples may be indicative of the limits of organism resolution of the method; while 16S sequence analysis is the prevailing molecular standard for determining phylogenetic relatedness, it may not be sufficient in all cases for strain-level identification, nor will it predict the presence of organisms expressing virulence factors. On the other hand, this method is compatible with rapid analysis of other conserved genomic regions and should be extendable to viruses (Scaramozzino et al. 2001). Finally, 16S rRNA sequences remain the largest dataset of gene-specific sequences, and, thus far, have proven as good as or better than other conserved genomic regions in determining relatedness.

The methods presented here are suitable for high-throughput identification of any materials enriched for a single dominant organism such as a weaponized biomaterial. Most importantly, it was shown that there is no fundamental limitation of the database searching technique or misidentification due to cleavage product mass degeneracies, even when very large databases of masses are used. Furthermore, due to the rapidity of MS acquisition and the ever-increasing amount of publicly available genomic information, other conserved sequence regions could be analyzed with little time penalty, thereby resolving any ambiguities among a particular group of organisms.

In general, segmentation of the analysis of the 16S gene into universally amplifiable subregions ultimately yields spectra of manageable complexity for accurate identification. With U and C residues better differentiated by 100% amino–allyl U substitution, acquired mass spectra can be "centroided" with increased confidence that strain-distinguishing masses differing by only a U or C residue are not convolved. The resulting spectra approximate a high-resolution "bacterial barcode" of minimal data and maximum information content.

Overall, these results indicate that complete cleavage after just one base should provide at least genus-level resolution of most bacteria, and that species- or strain-level identification may be achieved for some organisms using only the presented sequence regions of 16S rRNA. This can be improved by transcription and cleavage of the antisense strand, and/or cleavage after an alternative or additional base.

4.3.2 DNA Resequencing

The most common mutation screening technologies, like denaturing gradient HPLC, reveal information on the presence or absence of a sequence change without exact localization and characterization, and other methods like hybridization to oligonucleotide arrays and capillary sequencing locate the exact nature of sequence variations, but sometimes suffer from errors and are not easily ruggedized for field applications.

MALDI-TOF-MS has been used for resequencing of DNA by base-specific cleavage and MS analysis to identify DNA sequence changes on the level known as single-nucleotide polymorphism (SNP). Resequencing by base-specific cleavage and MALDI-TOF-MS on average facilitates the detection of >99.0% of all homozygous base-pair changes in 500 bp target regions. The technology has proven to be successful in the discovery of SNPs in human disease candidate genes, and signature sequence-based bacterial identification of pathogens (von Wintzingerode et al. 2002; Lefmann et al. 2004) as described in the previous section. Compared to earlier approaches, this new method increases the screened target region length between 5-fold and 10-fold and thus represents a major advance in establishing MALDI-TOF-MS as a resequencing tool. Overall, Honish et al. (2004) demonstrated that comparative sequencing by base-specific cleavage and MALDI-TOF-MS is an automated, fast, and highly accurate alternative to capillary sequencing (Honisch et al. 2004).

4.3.3 Triangulation Identification for Genetic Evaluation of Risk

Standard PCR-based tests, although can be performed in a relatively rapid manner, are failing to detect new or genetically modified pathogens. Therefore, Steven

Hofstadler and colleagues at Ibis Therapeutics and Science Applications International Corp. (with funding from the U.S. DOD DARPA) have developed a new strategy for identifying both known and previously uncharacterized pathogens (Hofstadler et al. 2005). They named this approach Triangulation Identification for Genetic Evaluation of Risk (TIGER). TIGER can be used to identify a wide range of organisms, such as viruses, bacteria, fungi, and parasitic protozoa.

The conventional methods use PCR primers to amplify species-specific regions of organisms' genomes and produce only true/false results. Hence, many iterations must be run. In case of TIGER, the diagnostic power comes from the use of broad-range primers, that is, primers that hybridize to highly conserved regions of the genome but that flank variable regions that have large differences in base composition. Therefore, the basis of TIGER is the use of these common, conserved features as anchors for broad-range PCR priming to generate amplicons from all organisms in an environmental or clinical sample without prejudice. For example, although a particular set of primers may bind to the genomes of all *Streptococcus* species, the actual PCR products will differ in a species- or even strain-specific manner. These differences are recognized by a MS as changes in the base composition that are expressed as different amounts of bases (A, T, G, and C) per amplicon. Although sequence information is not obtained by this method, the base composition from multiple primer pairs is used to "triangulate" the identity of the organism present in the sample (Hofstadler et al. 2005). If an unexpected composition is discovered, the researchers map the new signature in 3D or 4D space along axes representing each nucleotide. Base composition signatures for other organisms are also plotted. From this, a phylogenetic tree of relatedness can be constructed. Furthermore, TIGER can detect mixtures of organisms in the same sample by using base composition tags as signatures.

The process begins with the extraction of all nucleic acids present in a sample. The resulting nucleic acid material is divided and used as input to a number of PCR reactions, each with a single broad-range primer pair used for the amplification reactions (Figure 4.9). The PCR reactions take place in a 96- or 384-well format, and the resulting amplicons are purified for electrospraying into a MS. The resulting spectral signals are processed to produce a list of all detected masses, and the resulting masses are then converted into an unambiguous base composition of adenosines, guanosines, cytidines, and thymidines. For instance, as shown in Figure 4.10, the electrospray conditions separate the sense and antisense strands of the PCR products obtained from the severe acute respiratory syndrome (SARS) isolate. Because, each of the strands is present in multiple charge states; therefore two series of ions are observed, which represent ionic species with the same mass and different number of negative charges. By examining the isotope envelope of the $(M-27H+)^{27-}$ species, the derived molecular masses for the amplicon strands were determined as 27,298.518 ($\pm$0.03) Da and 27,125.542 ($\pm$0.03) Da. These masses correspond to an unambiguous base composition of A27G19C14T28/A28G14C19T27 for the double-stranded amplicon (Sampath et al. 2005).

In general, although the TIGER process depends on a PCR-based set of reactions, it is the unique set of primers and the unique form of the product analysis that allows detecting and identifying microorganisms down to a strain level. Moreover, the use of primer pairs that can each prime the production of amplicons from hundreds, if

Primer #	Gene target	Bacterial target	Primer specificity
346, 347, 348, 361	16S rDNA	All	Universally conserved ribosomal genes
349, 360	23S rDNA		
354	RNA polymerase, β subunit (rpoC)	Bacteroidetes, Fusobacteria, Spirochaetes, Proteo, Bacilli	Division-wide housekeeping genes
358	Valyl-tRNA synthetase (valS)	Proteobacteria (γ: Enterobacteria)	
359	RNA polymerase, β subunit (rpoB)	Proteobacteria (γ: Enterobacteria)	
362	RNA polymerase, β subunit (rpoB)	Proteobacteria (α, β)	
363	RNA polymerase, β subunit (rpoB)	Proteobacteria (β, γ)	
367	Elongation factor EF–Tu (tufB)	Proteobacteria (β)	
356, 449	Ribosomal protein L2 (rplB)	Clostridia. Fusobacteria Bacilli, Proteobacteria (ε)	
352	Protein chain initiation factor (infB)	Bacilli	
355	Spore protein (sspE)	*Bacillus cereus* clade	Clade-specific genes

350	Capsule biosynthesis protein (capC)	*Bacillus anthracis* pXO2 plasmid	Virulence plasmid
351	Adenylate cyclase (cyaA)	*Bacillus anthracis* pXO1 plasmid	Virulence plasmid
353	Lethal factor subunit (lef)	*Bacillus anthracis* pXO1 plasmid	Virulence plasmid

(The first block is bracketed and labeled "Survey"; the second block is bracketed and labeled "Drill-down".)

FIGURE 4.9 Bacterial primers for surveillance of BWAs—anthrax confirmation. (From Hofstadler, S.A. et al., *Int. J. Mass Spectrometry*, 242(1), 23, 2005.)

not thousands, of organisms removes the constraint of knowing exactly what one is looking for. In addition, the use of the MS instrument allows the detection of all the amplicons produced simultaneously, without the need for pre-separation of the components.

A simple way to visualize the information content in the base composition derived from a single primer pair is illustrated in Figure 4.11. In this case, a region coding a protein from the large ribosomal subunit (*rplB*) was amplified from multiple bacterial species (Ecker et al. 2006). The base composition results obtained during PCR are shown in a pseudo 3D coordinate system, where the number of A's, C's, and G's is plotted on the y-, x-, and z-axis, respectively. Also plotted (as colored spheres) are the base compositions one would expect to obtain (based on in silico PCR) from the known nucleotide sequences of multiple bacterial strains from various clades. The plot shows that this primer clearly differentiates *B. anthracis* from other bacterial pathogens shown on this plot such as *Streptococcus pneumoniae*, *S. aureus*, and *Streptococcus pyogenes* (and thousands of other pathogens not shown). Nevertheless, the final confirmation can be achieved by "drilling down" with a set of primers targeting sequences associated with the tripartite toxin and poly-D-glutamic acid capsule located on plasmids pX01 and pX02 (Figure 4.9).

The attractive feature of the TIGER is an exact identification that can be made not only by matching base compositions from several primer pairs to those stored in a large database, but also its capabilities of assigning identification to an unknown organism. This is made possible by the examination of the near-neighbor

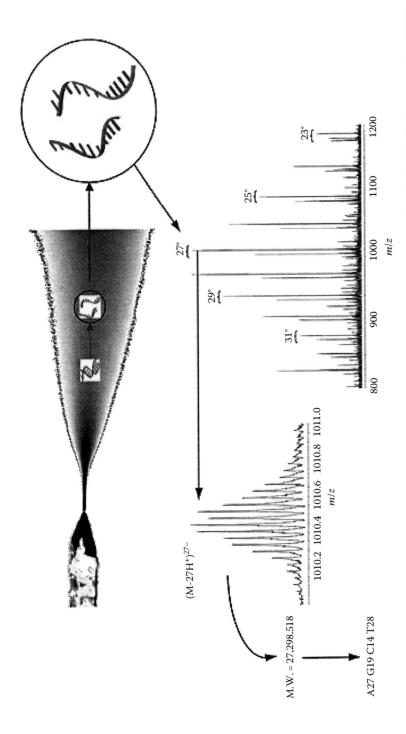

FIGURE 4.10 ESI-FTICR mass spectrum of the PCR amplicons from the SARS-associated coronavirus obtained with the propynylated RNA-dependent RNA polymerase primer pairs. (From Sampath, R. et al., *Emerg. Infect. Dis.*, 11(3), 373, 2005.)

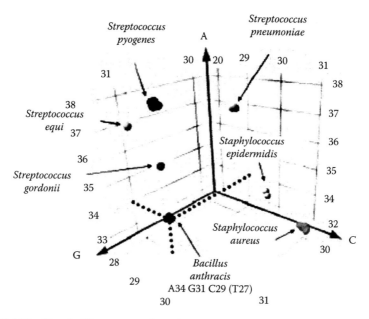

FIGURE 4.11 Pseudo 3D representation of the "base composition space" associated with the amplification of a region of *rplB* from multiple bacterial species. Note that the MS-derived base composition A34G31C29T27 clearly distinguishes *B. anthracis* from numerous other bacterial species. (From Ecker, D.J. et al., *J. Assoc. Lab. Automation*, 11(6), 341, 2006.)

relationships of the base compositions of the resulting amplicons and can be used to find closely related organisms and to assign a tentative identification to an unknown organism. TIGER can also be used for genotyping organisms by using a different set of primers and organizing organisms by the fine differences in small regions of their genomes.

Overall, MS methods are very sensitive and can measure the weight and determine the base composition from small quantities of nucleic acids in a complex mixture in a high-throughput manner. Hence, many laboratories are developing the next versions which are being designed as benchtop systems suitable in size for clinical diagnostics laboratories. Biowarfare agents, infectious disease epidemics, food contamination, and human forensics are just a few of the applications for the new methods.

4.4 DISCUSSION

What has taken place is a convergence of methods that seek genomic information and accuracy from MS methods. The convergence seeks a single method that can take genomic information from an unknown sample and associate it with the genomic information in the public databases to correctly produce an accurate identification. As the amount of genomic information has increased, this process has become a mathematical means that can process the vast information available from the public sources and then associate this information with data collected from samples from different environments such as plants, animals, food, etc. It is now possible to accomplish with software in a few minutes much of that which took long periods of wet work in the laboratory.

The processes in this chapter all helped evolve this methodology into a simplified system of sample processing, hardware, and software to accomplish the goal of accurate microbe identification. Sample processing has become standardized and improvements in hardware have been consistent and the software to analyze their output as seen in Chapter 6 is now capable of identify bacteria, viruses, and fungi with genomic accuracy.

5 Collection and Processing of Microbial Samples

Samir V. Deshpande

CONTENTS

5.1   INTRODUCTION

One of the main issues of any biological detection method is how do you collect the microbes. In the case of fungi, you can generally see the fruiting body and so you collect it into a pouch and move on; in the case of fungal hyphae, fungal spores, and the fungi imperfecti, also known as the Deuteromycota, the problem of collection becomes complex. Added to this, the bacteria and then the viruses and the sampling and collection methods suddenly become important to any method designed to detect and identify them.

Some of these methods are presented to provide a perspective to the issues of sample collection and presentation to the various MS methods. The good news is that with the introduction of MSP a simplified and one-step sample processing method has been developed for the purpose of detecting, identifying, and classification of microbes from complex biosamples.

5.2   SAMPLE COLLECTION METHODS

5.2.1   SAMPLE COLLECTION AND ENRICHMENT PROCEDURES

Samples analyzed using MS-based approaches for bacteria identification are usually processed in stationary laboratory settings; therefore, in majority of cases, researches follow standard procedures used for sample collection and pre-concentration and these will not be discussed here. However, those methods and technologies that have the potential for miniaturization and implementation in automated systems for analysis of environmental matrices are presented.

5.2.1.1   Aerosols

Air sampling for bioaerosols has been conducted for decades with classical monitoring that relies on collection using forced air samplers and analysis by either culture on artificial growth media or microscopy (Buttner et al. 2002). Aerosolized microbial agents of interest as potential biothreat agents tend to agglomerate as particles with an aerodynamic diameter of 0.5–10 μm and are of interest in average concentrations of ~15 agent containing particles (ACPs) per liter of air. Typically, aerosols over 10–20 μm fall out of the air rapidly and do not carry far from their origin, whereas the particles of interest will stay airborne for long periods.

One of the most common techniques used for fractionation of particles is the conventional impactor, which works by directing the particle-containing air through a nozzle onto a collection plate. Adjusting the nozzle, plate geometry, and flow parameters of the impactor determines the cut-point for the size that hits the plate and gets captured. A variation of the conventional impactor is the virtual impactor, which operates by directing the air stream from the nozzle to an opening with a restricted flow. Larger particles enter the opening, which forms a virtual surface, and become entrained in a minor flow of reduced velocity, while smaller particles follow the major flow. The virtual impactor has the benefit of concentrating particle quantity from low density in the high-volume flow to high density in the low-volume flow. The minor flow may capture 80% or 90% of the particles above the cut point and have a volume flow rate around one-tenth the total flow.

The smaller 1–10 µm particles can be collected by processing large volumes of air and passing the air through a filter or impinging the particles from the collected air into a liquid or semisolid sample (Stetzenbach et al. 2004). Impingement samplers that collect airborne cells into a liquid can collect particles over longer periods of time and are not limited by the types of analytical methods that can be used for detection of the collected microbes.

Collected air can also be passed through a porous filter. Filtration may desiccate vegetative cells, but the DNA and proteins from the cells would be preserved for PCR and proteomics type analysis. Filtration collectors are currently used in more than 30 major cities in the United States in the BioWatch biosurveillance program. Filters from these collectors are periodically removed and processed manually for the presence or absence of infectious agents by participating laboratories (Brown 2004).

The fieldable MS instruments developed for military applications were focused on analysis of air or biothreat agents from the surfaces. For example, the collection of particles for analysis by CBMS II was performed by using a novel, opposed jet virtual impactor (Romay et al. 2002) that draws air at ~300 L/min and concentrates them 150–270 times with efficiency of 50%–90%, in case of ambient particles in the size range of 2.3–8.4 µm. This particle-rich 1 L/min flow stream was directed to a pyrolysis tube and concentrated on the bottom for derivatization with a methanolic solution of TMAH for pyrolytic derivatization.

SCP Dynamics, Inc. has developed a batch-type aerosol-to-hydrosol transfer stage (AHTS) that is used with a virtual impactor characterized by an aerosol inlet flow rate of 1000 L/min and a coarse-particle flow rate of 20 L/min. The AHTS collects the particulate matter in the coarse-particle airflow into 40 mL of liquid or the SCP system to operate on a continuous flow basis. Two AHTS devices utilize circular jet impactors to deposit the particles from a 1 L/min aerosol flow rate into a liquid film that flows at a rate of 0.5 mL/min (Figure 5.1). The cut-points of the two devices are 0.8 and 2.5 µm AD. The liquid film forms on a porous surface through which the liquid is transpirated.

The Lawrence Livermore National Laboratory (Livermore, CA) has developed a high air volume to low liquid volume aerosol sampler that concentrates airborne materials from large volumes (2300 L/min) of air into a 4 mL liquid sample for subsequent automatic analysis. This sampler, part of the Advanced Pathogen Detection

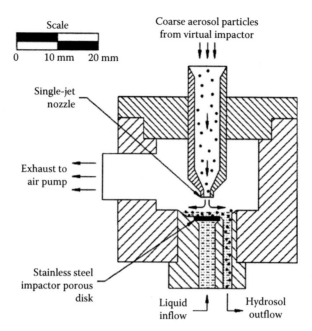

FIGURE 5.1   Cross-sectional view of AHTS. (From Phan, H.N. and McFarland, A.R., *Aerosol Sci. Technol.*, 38(4), 300, 2004.)

System, has been field tested for viable biothreat agents *B. anthracis* and *Y. pestis* in a biosafety level 3 facility (McBride et al. 2003).

Northrop Grumman (Arlington, VA) has incorporated the existing GeneXpert technology into its high-volume Biohazard Detection System for screening mail at U.S. Postal Service facilities. During mail operations, the Biohazard Detection System collects air samples directly above the cancellation equipment and concentrates air samples for 1 h by absorbing and concentrating airborne particles into a sterile water base. The fully automated system is a Cepheid GeneXpert module that identifies the presence of *B. anthracis* spores from air samples (Jaffer 2004).

A rapid, reagentless online analytical technique to sample and detect bioaerosols, at the individual particle level of resolution, called BAMS, is being developed at Lawrence Livermore National Laboratory for national security and public health applications that require detection of airborne bioagents such as vegetative bacteria, bacterial spores, viruses, and biological toxins (Fergenson et al. 2004; Russell et al. 2004, 2005). Air that may be contaminated with pathogenic bioaerosols is drawn into the device at a rate of about 1 L/min, and tracking lasers determine the velocities of any particle in the sample air. The BAMS system is based on a modified TSI-ATOFMS (Model 3800). In this system, aerosol is sampled at atmospheric background pressure via a 340 μm diameter nozzle into the MS to ~2.4 torr pressure. Two skimmers downstream collimate the particle beam and help reduce the pressure to ~10^{-4} torr in the particle sizing region. The velocity of the entrained aerosol is determined by measuring its transit time between two orthogonally arranged "sizing" laser (continuous wave 532 nm) beams. An orthogonal arrangement of the lasers ensures that only particles that travel straight down the center of the instrument are sized and counted. As particles cross the sizing laser beams, two scattered signals separated by particle transit time are produced and detected using photomultiplier tubes.

The TNO Prins Mauritius Laboratory, together with Delft University of Technology (The Netherlands), has developed a method for direct analysis of single aerosol particles by MALDI-MS. Therefore, the introduction of atmospheric particles into vacuum is achieved by passing air through a particle beam inlet system that consists of several pumping stages that simultaneously provide the particle enrichment in the aerosol stream. This aerodynamic lens system transmits particles with aerodynamic diameter in the range of 0.5–25 µm that were in-flight coated with a matrix.

The BioTOF-MS (Johns Hopkins University-Applied Physics Laboratory [JHU-APL]) features an air–air virtual impaction collection system that outputs a concentrated aerosol into a five-jet real impactor and subsequently deposits the particles onto VHS cassette magnetic recording tape (Anderson and Carlson 1999). Air drawn in through a plenum (Figure 5.2) flows through 6–10 nozzles. Each nozzle handles 3 L/min, with a pressure drop of 5.0 kPa. Particles accelerate in the nozzles through a 0.85 mm diameter acceleration jet to about 90 m/s and deposit by impaction on a section of tape located ~1 mm away. The tape is transferred for a deposition of the MALDI matrix and analyzed using the TOF-MS system.

5.2.1.2 Water

Microorganisms and their by-products in drinking water and recreational waters are usually too dilute for direct measurements. Water samples generally must be concentrated to obtain the number of target microorganisms necessary for microbiological analysis. Detection of microorganisms or their toxins from water samples is difficult because the recovery efficiency is poor and/or substances that may interfere with the detection technique are present. Target microorganisms must be isolated and concentrated by several orders of magnitude from the water samples. Concentration is most often performed using ultrafiltration in which parasites, bacteria, viruses, and potentially some high-molecular-weight biotoxins are retained on the filter (DiGiorgio et al. 2002). The material that has accumulated on the filter is removed, collected, and analyzed. Potential interferents such as salts, humic substances, and indigenous bacteria may also accumulate on the filter and inhibit rapid detection.

From the analytical standpoint, a major goal of sample collection methods is to provide enough material for successful analysis of targets at the level consistent with an infectious dose. To achieve this goal, more selective methods are used to enrich targeted bioagents in a sample. For example, the use of hollow-fiber flow field-flow fractionation (HF-FIFFF) was proposed to purify and fractionate different species of whole bacteria, thus allowing cells to separate from noncellular components that usually are present in the sample (Reschiglian et al. 2004). The application of HF FFF fractionation in combination with MALDI-MS analysis of samples provided a substantially improved quality of MALDI mass spectra acquired from whole bacteria.

The other approach aimed to enrich and separate target microbial cells prior to analysis uses dielectrophoresis. This technique takes advantage of the intrinsic

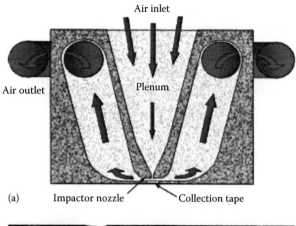

Air inlet

Air outlet   Plenum

(a)   Impactor nozzle   Collection tape

(b)

FIGURE 5.2 Aerosol impactor collector. (a) Movement of air flow and (b) the portable unit is linked by RS-422 to a remote notebook computer that provides monitoring and high-level commands. (From Anderson, C.A. and Carlson, M.A., *Johns Hopkins APL Tech. Digest*, 20(3), 352, 1999.)

dielectric properties of bioparticles to enable separation of particles in nonuniform electric fields. In dielectrophoresis, a nonuniform electric field consisting of positive and negative dielectrophoretic forces is generated by microelectrodes in a small chamber. Different bacterial cells and other components in the sample can be separated based on each particle's effective conductivity. Particles are released from regions near the electrodes when the dielectrophoretic response of each particle changes from attraction to repulsion from such regions. Different species of bacteria have different cell wall structures and compositions, and these differences give rise to large differences in particle conductivities. Changes in the physiological state of the cell will also affect particle conductivities. Dielectrophoresis has been incorporated into a microfabricated bioelectronic chip and used to separate *E. coli* from the sample (Cheng et al. 1998). Dielectrophoresis has also been used to clean target cells by removing inhibitors (Perch-Nielsen et al. 2003) and insulator-based dielectrophoresis has been used to isolate and trap *E. coli* and vegetative cells of *Bacillus* species into two distinct locations (Lay 2001).

The major problem associated with analytical approaches that rely on comparing microbial fingerprints for identification purposes is their inability to analyze microorganism mixtures or a single organism in a complex biological background. Therefore, one of the approaches to mixture analysis is using some type of capture molecules to selectively remove a targeted bacterium. The first application of affinity capture of whole bacterial cells coupled with MALDI analysis was demonstrated by Bundy and Fenselau (1999). They showed that lectins, i.e., protein moieties that bind carbohydrates, can be used to isolate gram-negative bacteria prior to mass analysis. The lectin concanavalin A was immobilized to a gold foil attached to a MALDI probe and used to retrieve *E. coli* from a buffer or a spiked sample of urine. MALDI-MS analysis confirmed that *E. coli* had been captured from the solution by identifying protein biomarkers in the resultant spectra. Further studies, reported by the same group, validated the ability of immobilized lectins to isolate bacteria and viruses from suspensions and even used immobilized carbohydrates to isolate bacteria by binding to lectins present on the surface of bacterial cells (Bundy and Fenselau 2001). Furthermore, a report by the same group showed that a microscope slide with the immobilized lectin motif could be used both for direct capturing of bacteria, after immersion in microbial suspensions, and could be directly inserted into MS for analysis as a modified MALDI plate (Alfonso and Fenselau 2003).

The Kent Voorhees group at the Colorado School of Mines showed that antibodies can also be used to purify and concentrate target biothreat agents. They used immunoaffinity-MALDI-MS techniques for bacterial isolation and detection by using antibodies immobilized on an active glass surface to retrieve and detect a targeted bacterial species from a biological mixture. Polyclonal antibodies, specific for all serotypes of *Staphylococcus*, were immobilized on glass. Glass plates with the immobilized antibody were allowed to incubate for 1 h in bacterial solutions containing only *S. aureus* or a mixture of *S. aureus* and *E. coli*. Upon incubation, glass plates were subjected to washings to remove unwanted material from the surface of the plate. The addition of trifluoroacetic acid solution followed by the deposition of a matrix (ferulic acid/formic acid [FA]/acetonitrile [ACN]/H$_2$O) and mass analysis revealed the capabilities to distinguish *Staphylococcus* species from a mixture of organisms. In the other format, magnetic particles (beads) with antibodies immobilized on their surfaces are used for binding to target cells, toxins, or other molecules found in samples (Madonna et al. 2001). The magnetized particles are then collected using a magnetic field. Sample debris and nontarget organisms and molecules are removed by washing. The particles are released when the magnetic field is removed. The remaining solution contains mainly concentrated cells or molecules of interest that have bound to the antibody attached to the magnetic particle. This antibody-based method of cell separation is referred to as immunomagnetic separation. *Listeria monocytogenes*, *Salmonella* spp., *E. coli* O157:H7, and *Cryptosporidium* spp. have been successfully isolated from water samples using immunomagnetic separation (Bukhari et al. 1998), and the method is a component of the U.S. Environmental Protection Agency (EPA) Method 1622 and Method 1623 for the recovery and identification of *Cryptosporidium* spp. and *Giardia* spp. in environmental waters (U.S. EPA 1999). Immunomagnetic separation is carried out using beads that are functionalized for covalent linkage to antibodies or peptides targeting biothreat agents, which are commercially available (Dynabeads) from Dynal Inc. (Oslo, Norway). Beads with linked protein A or G for noncovalent attachment of antibodies to the beads are also available.

Immunomagnetic separation has been used in a field-deployable automated electromagnetic flow cell fluidics system (Biodetection Enabling Analyte Delivery System [BEADS]) to separate and concentrate pathogenic bacterial cells as well as nucleic acid. The BEADS system engineered by Pacific Northwest National Laboratory (Richland, WA) was designed specifically for processing environmental and clinical samples prior to biodetection. Porous Ni foam was used to enhance the magnetic field gradient within the flow path so that the immunomagnetic separation particles could be immobilized throughout the fluid rather than at the tubing wall (Chandler et al. 2001; Chandler and Jarrell 2004).

Commercially available antibodies targeted against the O and K antigens were used for capturing the *E. coli* onto beads, and the recovered beads were used for direct PCR amplification and microarray detection (Chandler et al. 2001). The target microbial pathogens have also been recovered from a water sample using immunomagnetic separation (Enroth and Engstrand 1995) and this approach proved to be valuable for MALDI-MS-based detection of microbes. As long as antibodies or other affinity-based molecules are available for attachment to the beads, immunomagnetic separation can be a versatile technique potentially available for the purification of target biothreat agents and their products from heterogeneous sample matrices, followed by MS analysis. Furthermore, it should be pointed out that some methods developed for water analysis are also appropriate for analysis of a liquid suspension of particles concentrated from bioaerosols by using AHTS described earlier.

The use of immunomagnetic beads with antibodies to isolate bacteria from mixtures for MALDI-MS analysis was demonstrated by Madonna et al. (2001). In this approach, immunomagnetic beads with antibodies specific for the target biological agent are mixed with the sample for a short incubation period (minutes), then isolated by a magnet, and washed free of background constituents. Following the washing step, the beads are then resuspended into a smaller volume of water and applied to the MALDI sample probe. The deposited beads are covered with a matrix solution and mass analyzed. The resulting mass spectrum is then evaluated for the presence of specific biomarkers of the suspected organism. The results obtained with this method have shown that anti-*Salmonella* immunomagnetic beads could be used to ambiguously determine the presence of this organism in spiked samples of river water, urine, blood, and milk. They illustrate the powerful capabilities of combining immunoassays with MS.

In the more recent example of using magnetic particles for sample enrichment, Lin et al. (2005a) employed vancomycin-modified magnetic nanoparticles as affinity probes to selectively trap gram-positive pathogens from sample solutions and to isolate nanoparticles by applying a magnetic field. The isolated cells were characterized by MALDI-MS and the obtained spectra indicate that this approach effectively reduces the interference of protein and metabolite signals in the mass spectra of gram-positive bacteria because vancomycin has very high specificity for the D-Ala–D-Ala units of the cell walls. The lowest cell concentration detected for both *Staphylococcus saprophyticus* and *S. aureus* in a urine sample (3 mL) was ~7 × 10⁴ cfu/mL (Lin et al. 2005a).

Aptamers and peptide ligands are alternatives to antibodies that recognize a target by shape, not by sequence, and are generated using combinatorial methods. Aptamers have been used to detect the toxin ricin in a bead-based biochip sensor (Kirby et al. 2004) and cholera toxin, staphylococcal enterotoxin B, and *B. anthracis* spores in an electrochemiluminescence assay (Bruno 1999; Bruno and Kiel 2002) and could be easily adapted for MS-based analysis. A review by Breaker (2004) provides more detailed information on the characteristics and potential applications of aptamers and similar affinity probes.

Although conventional antibodies are the predominant affinity probe used in shape recognition-based technologies, mono- and divalent antibody fragments such as Fab′ and F(ab′)₂ as well as single-chain variable regions have also been explored to determine advantages in sensitivity, specificity, or durability compared to antibodies. Such fragments may also be modified using recombinant technology, thus making them more favorable as probes (Petrenko and Sorokulova 2004). Using this methodology, probes have been developed that bind to bacteria, viruses, and toxins, e.g., *C. difficile* toxin B, *B. melitensis*, vaccinia virus, and botulinum toxin (Emanuel et al. 2000).

Short recombinant peptide sequences have also been tested as capture and detection elements in biosensors. As with phage display-generated antibodies and fragments, these peptide ligands may be chemically synthesized or remain as a phage probe. Sequences that bind specifically to ricin, *B. anthracis*, and other *Bacillus* species spores (Turnbough 2003; Williams et al. 2003a; Brigati et al. 2004), staphylococcal enterotoxin B, and protein A of *S. aureus* (Mason et al. 2003) have been developed and tested using non-MS approaches; however, they could be easily coupled with MS-based methods as demonstrated earlier.

5.2.1.3 Food

A food sample submitted for analysis is usually blended or rinsed to provide suspended cells. The integrated system developed by the Food and Drug Administration for characterization of bacteria by MAB-Py-MS uses (1) automated, accelerated bacterial enrichment and selective enrichment, followed by (2) instrumental isolation out of liquid culture media of individual cells, and (3) automated cell washing and cell suspension normalization to standard optical density. The integrated system is comprised of (1) a robot for dispensing culture media, solvents, cell suspensions, or other liquid reagents into 96-well microtiter plates, (2) a centrifuge with a rotor for microtiter plates, (3) a UV/vis microtiter plate reader, and (4) a small microbiology incubator. In addition, the earlier system uses a flow cytometer equipped with scattering detectors and sorting option that can distinguish cell size and shape. It performs sorting of cells with a specified dimension into culture broth-filled wells for culturing. These subsystems, combined with automated analysis using an MAB-Py-MS technique, can give rapid distinction of bacteria at near strain level after a single working day's operation.

Obstacles to preparation of food samples for testing using rapid detection protocols include the presence of food components such as lipids, polysaccharides, and
salts, which may interfere with the detection of the target pathogens and decrease
sensitivity of detection for rapid analysis. Overall, the rapid and effective separation
and concentration of target pathogens remains a barrier to the rapid detection of
microorganisms in foods and many current methods still rely on the time-consuming
enrichment culture for the growth of the target microorganisms in a sample.

5.2.2 BACTERIOPHAGE AMPLIFICATION USED FOR BACTERIA DETECTION AND IDENTIFICATION

Bacteriophages are known to specifically infect bacteria even at the strain level and
have been used in the past for phage-typing characterization of bacteria. When bacteriophage attacks the bacterial cell, the virus takes over the cell biochemistry and
multiplies itself. Assuming that the number of virions that escape from one infected
bacterial cell may reach 10,000 and each contains many copies of the same protein
(e.g., 180 copies of a capsid protein in case of the MS2 virus that infects *E. coli*), a
single cell could be amplified by a factor of 1.8×10^6 (Madonna et al. 2003).

5.3 WHOLE-CELL PROCESSING METHODS FOR FINGERPRINTING MICROBIAL AGENTS

5.3.1 PYROLYSIS

Chemical characterization of microorganisms can be achieved by probing chemical composition of cells through direct pyrolysis, followed by analysis of thermal
fragmentation products (Snyder et al. 1994) or, alternatively, by thermolysis in the
presence of chemical derivatization agents. The latter mode of pyrolysis aims to
transform nonvolatile and/or polar macromolecular constituents of any cell into
lower polarity compounds, thus facilitating the efficient transfer of analytes to the
MS analyzer (Dworzanski and Meuzelaar 1988; Dworzanski et al. 1990).

With pyrolysis techniques, the sample pretreatment may be limited to placing a
microbial sample on the pyrolysis probe and allowing drying or a sample can be
additionally mixed with a derivatization agent before heating. Thermal degradation
of structural components of cells usually produces an extremely complex mixture
of fragments and, hence, makes their identification difficult. On the other hand, the
pyrolysis process performed in the presence of a methylating agent, e.g., TMAH,
gives very profound changes in the composition of pyrolyzates. For example, pyrolytic products of bacteria are dominated by the presence of fatty acid methyl esters.
Because these components were traditionally used for the characterization and
identification of bacteria, this approach was adopted for the detection and classification of biothreat agents by CBMS II. However, such approach requires an
additional subsystem designed for delivering methanolic solution of a methylation
agent (Barshick et al. 1999).

In the case of bacteria derived from food samples and analyzed by Py-MAB-
TOF-MS, an autosampler conjoined with an autoprobe was used, which is

engineered for unattended sample introduction, and programmed pyrolysis inside the MAB-TOF-MS instrument (Dephy Technologies, Montreal, Canada; Chapter 8.2) (Wilkes et al. 2005b).

5.3.2 WHOLE-CELL MALDI

The very nature and profuse range of proteins available for bacterial proteomics investigations inherently may provide many avenues for experimental irreproducibility of recorded spectra. It may seem difficult to provide conditions for satisfactory mass spectral reproducibility given the thousands of proteins at significant different concentration levels in microorganisms. However, as outlined later, different data analysis methods provide different tolerance and latitude in the experimental parameters to yield satisfactory and reproducible results. Nevertheless, pre-analysis sample preparation steps incorporate the most important elements influencing the quality and reproducibility of the spectra and include the type of matrix, the matrix solvent, and concentration of cells in the matrix, as well as the type and concentration of acid added to the matrix (Williams et al. 2003b).

Processing of bacterial samples is a very important issue, because the quality of spectra influences the integrity and robustness of any data analysis algorithm. The following presents sample and instrumental parameters that were investigated over the last decade to obtain high-quality and reproducible mass spectra of bacterial proteins.

Three general processing procedures are used to generate protein ions for subsequent characterization and identification of bacteria with MALDI-MS. The simplest of them uses a mixture of the bacterial sample with a matrix deposited onto a metal MALDI target (Wang et al. 1998). In this whole-cell analysis procedure, any deliberate steps to break open or fragment the exterior cell walls are avoided in part due to logistic and procedural concerns. The second method consists of suspending the whole cells in a solvent capable of solubilizing or extracting protein species from the bacterial sample (Hettick et al. 2004). Different types and amounts of proteins are extracted depending on the polarity of the solvent and the presence of additives and a portion of the protein extract is mixed with matrix and analyzed by the LDI-MS. The efficiency of protein extraction was investigated with respect to different solvents (Williams et al. 2003b), pH (Domin et al. 1999), salt content (Ruelle et al. 2004), detergent (Bornsen et al. 1997), and other additives (Krishnamurthy et al. 1996).

The third method uses lysis techniques (Birmingham et al. 1999) to deliberately break open or fragment the bacterial cell, and this allows straightforward solvent extraction of cellular proteins. A protein extract is mixed with a matrix and the mixture is analyzed by MALDI-MS. An important consideration in the processing of a bacterial sample is whether it is in a spore or vegetative state, because spores are characterized by a relatively hard, resistant outer cell wall. Generally, fewer numbers of proteins are released and observed in spore mass spectra than from gram-positive and gram-negative vegetative bacteria. Thus, spores generally require extended or more strenuous processing conditions in order to realize a sufficient number of protein peaks for statistical purposes in the characterization and differentiation of the organism.

In addition to conditions used to extract and mass analyze protein species, bacterial growth parameters are important for the reproducible generation of protein mass spectra and should be carefully controlled. These include different growth media (Ruelle et al. 2004) and growth conditions, e.g., flask vs. test tube, aerated liquid broth vs. solid agar plate growth, and growth time or phase of the organism (Arnold et al. 1999).

Vortex or mixing time of the bacteria (Wang et al. 1998) in the extraction solvent can produce a differential amount of proteins in the resulting mass spectrum. This is particularly true when comparing the spore vs. vegetative state of a gram-positive organism. The matrix itself has been intensely studied with regard to its efficiency in transferring energy to the many protein analytes because the complexity of the biological analyte and chemical constituents in the matrix (Williams et al. 2003b) significantly determines the extent of protein ionization, magnitude, and dynamic mass range of the ions in the mass spectrum. Protein extraction efficiency can be modified by type, number, and amount of other material in the matrix. These materials include additional matrix substances, complexity, and concentration of the many proteins in the extract, presence of salts, detergents, acid addition, residual growth media (Wang et al. 1998), and proteins from more than one organism in the original extract. The addition of a protein mass standard as an internal calibrant may affect the presence/absence of certain organism protein ions (Gantt et al. 1999).

A number of reports focus on the laser power, intensity, and laser spot size in the efficient desorption and ionization phenomena that fundamentally characterize the MALDI process (Wunschel et al. 2005). The influence of these laser parameters on the matrix and the complex biological analyte and extraneous interference milieu interact in such a way as to produce a MALDI mass spectrum that represents a percentage of the protein species in the sample spot.

Researchers at Johns Hopkins University (McLoughlin et al. 1999) have addressed the problem of sample collection by combining their instrument with an aerosol impactor. The material was deposited on a video tape, which is periodically transported into the MS for analysis. This configuration reduces the total analysis time to 10 to 15 min, which is still relatively long for warning detection purposes.

5.3.3 DESI-MS OF INTACT BACTERIA

A technique called DESI combines features of ESI and desorption ionization methods (Cooks et al. 2006) by directing a stream of electrosprayed droplets of a suitable solvent onto a surface to be analyzed.

DESI minimizes the requirements for sample preparation by enabling investigation of samples with bioagents in their native environment or on the surface of a filter used for sample concentration. In short, the microbial constituents are extracted by charged droplets produced from the electrospray and are directed by a high-velocity gas jet to the analyzed surface. The charged impacting droplets dissolve microbial constituents and new secondary droplets are ejected from the surface and directed into an atmospheric pressure inlet of a standard commercial MS for mass analysis. The fact that matrix solutions are not required for ion generation is considered as a significant advantage over the MALDI technique because it simplifies sample

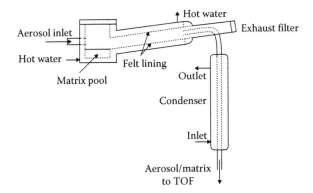

FIGURE 5.3 Evaporation/condensation cell to cover the aerosol particle with matrix.

processing. The two methods, however, are complementary in that MALDI is primarily suited for detection of large molecules such as peptides and proteins, and DESI is well suited for detection of small molecules such as lipids and metabolites.

5.3.4 SINGLE-PARTICLE LDI

Researchers at the Delft University of Technology and TNO Defence, Security and Safety developed a technique that combines fluorescence preselection and MALDI-TOF-MS to obtain mass spectra from single biological aerosol particles (Marijnissen et al. 1988; Kievit et al. 1996; Stowers et al. 2000; Van Wuijckhuijse et al. 2005). Mass spectra from a single or a few individual aerosol particles can be obtained within seconds, enabling rapid identification of the aerosol material. The system incorporates an evaporation/condensation flow cell that allows in-line matrix coating of the aerosol particles (Figure 5.3). The coated particles are separated from the surrounding air using aerosol beam techniques. The passage through the focal point of a UV-laser induces fluorescence in biological materials. This effect can be used for preselection, which enables the preferential analysis of aerosol particles of biological origin. Detection and sizing is achieved by collecting the light scattered when the particles pass through the focal points of a second laser beam. This event also triggers the desorption laser. The matrix of UV-absorbing material ensures efficient ion formation, and the resulting mass spectra can be used to identify the aerosol material.

5.4 CELL LYSIS AND EXTRACTION OF SPECIFIC CELLULAR COMPONENTS

5.4.1 BACTERIAL CELL LYSIS METHODS

Successful detection and characterization of microbial agents using a variety of analytical techniques is based on the interrogation of nucleic acids, proteins, lipids, and other cellular components. Therefore, lysis of cellular microorganisms usually represents the first step aimed at releasing these constituents using physical or chemical means for rupturing membranes/envelopes surrounding a cytoplasm. Many techniques are available for the disruption, which utilize both physical and chemical phenomena for an efficient lysis.

5.4.1.1 Physical Lysis Techniques

Physical lysis has been the method of choice for cell disruption; however, it often requires expensive, cumbersome equipment and involves protocols that can be difficult to repeat due to variability in the apparatus. The most reliable methods in this category include ultrasonication, French press, pressure cycling, bead mills, and thermal lysis. Recently, new methods for lysis of microbial cells directly on the chip were investigated and applied to pathogen detection systems.

5.4.1.1.1 Ultrasonication

Cell membrane disruption by sonication is directed by ultrasound-induced cavitation. Ultrasonic waves propagate in liquid media as pressure waves that alternatively expand and contract and in so doing create microbubbles or "cavities." Collapse of these cavities can produce extreme shear forces with the ability to disrupt membranes. The method uses pulsed, high-frequency sound waves to agitate and lyse bacteria and spores. The sound waves are delivered using an apparatus with a vibrating probe that is immersed in the liquid cell suspension. Mechanical energy from the probe initiates the formation of microscopic vapor bubbles that form momentarily and implode, causing shock waves to radiate through a sample. To prevent excessive heating, ultrasonic treatment is applied in multiple short bursts to a sample immersed in an ice bath. Using a sonicator tip with small volume of biological sample usually requires insertion of the tip deep in the solution vial to avoid foam formation. Also, addition of thiol-containing compounds, i.e., dithiotreitol (DTT) or cysteine could minimize the oxidative reaction that resulted from radical formation during the sonication process. Common sonication devices include a sonication bath, ultra tip-sonicator, and miniaturized sonicators (Figure 5.4) such as a mini-sonicator developed by Belgrader et al. (1999).

FIGURE 5.4   Flow cell sonicator. (From Belgrader, P. et al., *Anal. Chem.*, 71(19), 4232, 1999.)

5.4.1.1.2   French Press

French press lysis occurs when the cell suspension is pressed through a small capillary. The pressure difference between the chamber and the capillary ruptures the cell. French press is categorized as a high-pressure homogenizer, which utilizes a motor-driven piston inside a steel cylinder to develop pressures up to 40,000 psi. Pressurized sample suspensions are forced through a needle valve at a rate of about 1 mL/min. Because the process generates heat, the sample, piston, and cylinder are usually pre-cooled. Typical pressures used to disrupt bacteria are 8,000–10,000 psi and several passes through the press may be required for high efficiency of disruption. Generally, the higher the pressure, the fewer the passes required to rupture the cell wall. The operating parameters which effect the efficiency of French press technique include pressure, temperature, number of passes, valve and impingement design, and flow rate.

The rate of cell disruption by French press is roughly proportional to the third power of the turbulent velocity of the product flowing through the homogenizer channel, which in turn is directly proportional to the applied pressure. Therefore, the higher the pressure, the higher the efficiency of disruption per pass through the machine.

French press has long been considered the best available means to mechanically disrupt nonfilamentous microorganisms on a large scale. The supremacy of French press for disruption of microorganisms is now being challenged by bead mill homogenizers. Still, in terms of throughput, the industrial models of French press homogenizers outperform bead mills.

5.4.1.1.3   Pressure Cycling

The pressure cycling technique (PCT) is based on exposing the biological sample present in container to a cyclic hydrostatic pressure from low to high levels, e.g., from ambient to 235 MPa. PCT is capable of processing both solid tissues and liquid cultured cells, including bacterial spores and other hard-to-lyse materials. Several cycles are employed and 3–5 one-minute cycles have been found to be sufficient for processing most sample types. Lysed samples are then ready for extraction of nucleic acids and/or proteins for analysis using MS or other bioanalytical techniques (Tao et al. 2003).

5.4.1.1.4   Thermal Lysis

Thermal lysis is a term that is used to indicate the application of heat to lyse cells. This includes the freeze/thaw approach and lysis at temperatures approaching the solvent boiling point. Thermal lysis using the freeze/thaw method is commonly used to lyse bacterial cells. The technique involves freezing a cell suspension in a dry ice/ethanol bath or freezer and then thawing the material at room temperature or elevated ones (37°C–100°C). This method of lysis causes cells to swell and ultimately break as ice crystals form during the freezing process and then contract during thawing. Multiple cycles of heating are usually employed for efficient lysis, and the process can be quite lengthy depending on the physical dimension of the heating apparatus and sample volume (Pietsch et al. 2001). Although the freeze/thaw technique is still useful in some laboratory applications, e.g., to release recombinant proteins located in the cytoplasm of bacteria, it is not suitable for fast sample processing.

To increase the speed and efficiency of lysis, various thermal treatments were proposed that include microwave heating, sonication, and thermal shocks caused by fast heating to temperatures approaching water boiling point. For example, Yeung et al. (2006) used a chamber maintained at 90°C for complete lysis of both gram-positive and gram-negative bacteria using a 5 min long cell residence time in the hot zone. This process was performed directly on the chip. Furthermore, at the same time denaturation of DNAs and proteins takes place, which can be immediately processed for pathogen identification.

5.4.1.1.5 Laser-Irradiated Magnetic Bead System

Lee et al. (2006) demonstrated rapid lysis of bacterial cells using both thermal and mechanical lysis directly on the chip through a combination of the laser irradiation and agitation with magnetic beads. For this purpose, a 808 nm small-size laser diode and carboxyl-terminated magnetic beads in a microchip chamber were used. This provided efficient cell lysis and DNA isolation for various pathogen detections. They also performed this type of lysis in a disposable microchip as a portable sample preparation device with vibrating magnetic beads present in cell suspension that caused rapid cell lysis and DNA purification when the magnetic beads were irradiated by a laser beam. They demonstrated a single pulse of 40 s lysed pathogens including *E. coli* and gram-positive bacterial cells (*Streptococcus mutans, S. epidermidis*). They further demonstrated that the real-time pathogen detection was performed with pre-mixed PCR reagents in a real-time PCR machine using the same microchip, after laser irradiation in a hand-held device equipped with a small laser diode. These results suggest that the new sample preparation method is well suited to be integrated into lab-on-a-chip application of the pathogen detection system and could be applied for sample processing prior to MS analysis.

5.4.1.2 Reagent Lysis

Reagent cell lysis is often applied to various cell types to achieve both cell rapture and a selective way of extracting certain cellular components. There are two major types of reagents used to lyse microbial samples, namely chemical and biochemical reagents. The following sections provide a brief description of both approaches.

5.4.1.2.1 Chemical Lysis

The original goal of lysing cells using chemical reagents was to preserve the cell substantially intact even after the release of its contents. This facilitates separation of the cell debris from the supernatant. However, the compatibility of the chemical reagent with the analytical technique must be considered. In the case of protein extraction, the removal of lysed DNA from the lysates is necessary to eliminate the viscosity issue that might occur during protein purification. Chemical lysis usually involves the addition of chelating and chaotropic agents, and/or detergents. Chelating agent in addition to bacterial cells will diminish the cell wall permeability barrier, while chaotropic agents, i.e., guanidine, ethanol, and urea, will weaken hydrogen bonds and act as denaturizing agents that destroy higher-order structure of macromolecules and cause their denaturation. Unfortunately, these agents are usually employed at high concentrations, which make them unsuitable for large-scale lysis processes. Further, ionic and non-ionic (Triton X-100, Brij, Duponal) detergents have been used to permeabilize bacterial inner cell membranes. Various combinations and types of detergents are used to release specific cellular components, such as periplasmic or cytoplasmic proteins.

In recent years, detergent-based lysis has become very popular due to ease of use, low cost, and efficient protocols. Several vendors offer several detergent-based reagents for the preparation of whole and fractionated cell lysates that are faster and more convenient than traditional lysis methods. Furthermore, a new generation of detergents has been developed for MS applications.

5.4.1.2.2 Biochemical Lysis

Biochemical cell lysis employs certain enzymes and proteins. Most notable biochemical lysis is by using lysozyme to hydrolyze $\beta1\rightarrow4$ glycosidic linkages in the peptidoglycan of bacterial cell walls. This approach is very effective for gram-positive bacterial cells to lyse rapidly. However, gram-negative bacterial cells require pretreatment step to allow lysozyme to access the cell wall. Such a process requires the addition of chelating reagents such as ethylenediaminetetraacetic acid (EDTA). Many other biochemical reagents have been reported; however, they are of a limited value for field or mobile laboratory applications.

5.4.2 EXTRACTION OF SPECIFIC MICROBIAL CONSTITUENTS AND SAMPLE INTRODUCTION FOR MS ANALYSIS

Conventional culture and staining techniques are currently the gold standard for isolation, detection, and identification of target biothreat agents. However, the use of culture enrichment and selection results in lengthy assays, which can take days for preliminary results. Rapid detection methods replace the selective and differential culturing steps with DNA hybridization, nucleic acid amplification, antibody agglutination, enzyme immunoassays, and MS-based techniques. The majority of these rapid detection methods are suitable only when the biothreat agent is present in sufficient number and/or in the absence of interfering substances.

In most cases, rapid detection methods require steps to concentrate the target biothreat agent and/or purify the target analyte from the sample matrix prior to rapid detection. For example, PCR and nucleic acid sequence-based amplification may enrich a single, specific DNA or RNA sequence up to 10^6-fold in 20 min to a few hours and theoretically has the sensitivity of a single bacterial cell. However, substances such as bile salts, polysaccharides, heme, and humic acids in sample matrices inhibit enzymatic reactions required for nucleic acid amplification (Rådstrom et al. 2004). In general, low levels of target analyte in samples require concentration and/or cultural enrichments to provide sufficient target for amplification. Although many MS-based methods of low specificity do not require sample preprocessing; nevertheless, the more advanced MS methods, which rely on revealing sequence information of nucleic acids or proteins, usually need sample processing before it can be introduced into a NMS.

5.4.2.1 Methods for Sample Processing

Ideally the method to separate, concentrate, and purify the target biothreat agent should be universal, utilizable for all samples for all types of target analytes. In addition, the sample preparation method should be capable of rapidly removing the sample matrix that could inhibit detection capabilities and concentrating the analyte. Therefore, the preparation procedure is usually limited to specific types of samples (e.g., aerosol) and is generally time-consuming. Nevertheless, many automated sample preparation (ASP) methods are currently under investigation. These methods include chemical, physical, and biological manipulation of the sample and many of these methods have been developed for testing food for the presence of pathogens (Benoit and Donahue 2003; Rådstrom et al. 2004; Stevens and Jaykus 2004). In this section, we will focus on rapid and/or ASP methods for various liquids and aerosols and will include methods that are commonly used for detection of biothreat agents.

The most commonly used methods to recover and/or concentrate targeted species include centrifugation and diverse filtration procedures. Centrifugation is a method that has conventionally been used to concentrate and recover microorganisms from liquid samples. Moreover, undesired sample debris may also be concentrated during the process and the centrifugation process cannot be easily automated. For example, *E. coli* O157:H7 cells have been recovered directly from a ground beef/buffer suspension by a 5 min differential centrifugation step that separated the suspension into three distinct layers (DeMarco and Lim 2002). The middle layer containing the majority of the target cells was used for rapid detection of *E. coli* O157:H7. Buoyant density gradient centrifugation has been used to separate and concentrate *Y. enterocolitica* in meat fluids from pork (Wolffs et al. 2004). This centrifugation procedure removed dead cells, and the concentrated samples contained only viable cells that were then used directly for the PCR-based amplification of targeted sequences.

Filtration can be used to separate microorganisms on the basis of cell size. Although liquid samples can be rapidly forced through filters of different pore sizes, sample debris can clog filters and retain bacteria. Removal of bacteria from filters following filtration can also be difficult. Special filters (Whatman, Springfield, KY) have been developed for rapid isolation of nucleic acids from environmental, clinical, or food samples. Samples are added directly to the filter. The filter can then be washed and the nucleic acid remains bound to the filter. The filter is then ready for processing or long-term storage at room temperature. Such filters have been used for the detection of *B. subtilis*, *B. cereus*, and *B. megaterium* spores using nested PCR. The reported sensitivity of this method is 53 spores in the first round of PCR, and 5 spores after the second nested PCR round (Lampel et al. 2004). In addition, centrifugal devices, which take advantage of centrifugal forces to speed-up membrane filtration, solid-phase extraction, and gel permeation chromatography, are commonly used.

5.4.2.1.1 Lipids

Among many types of lipid species composing microbial cells, fatty acid profiling of cultured bacteria is most frequently used for the identification of bacteria to the species level, thus providing useful taxonomic information. To this aim, simple and reliable methods have been developed for determination of whole-cell fatty acids, which are usually analyzed in the form of methyl esters by a GC-MS technique. Nevertheless, culturing conditions may have a dramatic influence on the results obtained; therefore, both standardized culturing and sample-processing methods are usually implemented.

Fatty acids are mainly present in bacterial lipids in the form of ester- or amide-bonded moieties of phospholipids, glycolipids, proteins, and some other components of bacterial cell membranes and walls. Although the most frequently applied methods are based on the conversion of acids to methyl esters, other derivatives are also used, for instance, pentafluorobenzyl or *tert*-butyl-dimethylsilyl esters. Methyl esters are usually prepared by acid methanolysis, alkaline methanolysis, or saponification with dilute alkali in aqueous methanol of whole cells, followed by sample clean-up and esterification of acids (Jantzen 1984).

For microbial characterization by phospholipid profiling, a procedure based on simple extraction of polar lipids with a chloroform/methanol mixture is usually applied (Bligh and Dyer 1959). The extract is centrifuged and a lower phase containing phospholipids is collected for direct ESI-MS analysis. However, a chloroform/methanol procedure could be replaced by other solvent systems (Vaidyanathan et al. 2002) or a supercritical fluid extraction, thus reducing costs, increasing the speed, and eliminating waste solvents (Pinkston et al. 1991).

Vaidyanathan et al. (2002) reported that unfractionated ACN/0.2% FA extracts obtained from bacterial cells could be directly analyzed by flow injection ESI-MS (FI-ESI-MS) for a rapid, reproducible, and high-throughput bacterial identification. Five bacterial strains (two *E. coli*, two *Bacillus* spp., and one *Brevibacillus laterosporus*) were studied in this investigation and the cell-free extracts were sequentially injected into a solvent flow stream that was electrosprayed into the MS. The spectra produced contained sufficient information for discriminating between the bacteria. They showed that acquired spectra were dominated by phospholipids; however, glycolipids and proteins were also contributing to the spectral information. Furthermore, analysis of the extracting solvent components showed that ACN contributes most significantly to the extraction process and hence to the information content of the spectra.

5.4.2.1.2 Carbohydrates

In certain instances, carbohydrate profiles can be more informative than profiles of fatty acid methyl esters. For example, differences in carbohydrate patterns could be used for discriminating *B. anthracis* and *B. cereus* strains (Fox et al. 1993). Carbohydrate markers for bacteria were usually extracted from whole-cell acid hydrolyzates aimed to release monosaccharides from macromolecules. The released sugars are analyzed using GC or LC coupled with MS. In both types of analysis, samples are hydrolyzed with sulfuric acid, neutralized, cleaned, and subjected to chromatographic analysis. For example, Wunschel et al. (1997) performed hydrolysis of whole bacterial cells with diluted sulfuric acid which was next neutralized with a mixture of *N,N'*-dioctylmethylamine/chloroform. The aqueous phase was freed from hydrophobic and cationic contaminants by solid-phase extraction with C-18 and SCX resins and injected into an LC-MS/MS system. Sugars were separated on a

4 mm i.d. Carbopack pellicular anion-exchange column using a concentration gradient of a sodium hydroxide and ammonium acetate solution as an eluent. During a 45 min long analysis, the effluent from the 4 mm i.d. column was neutralized using an anion suppressor run in the external water mode. In this approach, water regenerates free hydrogen ions through water hydrolysis using 300 mA current, and sodium hydroxide is replaced by water and sodium acetate by acetic acid.

The sample preparation for a GC procedure is more demanding and relies on the so-called alditol acetate method for sugar analysis. In this method, aldehyde groups are reduced with borohydride to eliminate anomeric centers and all hydroxyl and amino groups are next acetylated. Because both post-derivatization and pre-derivatization clean-ups are required, the whole procedure is labor-intensive and difficult for automation. Efforts to develop simpler methods were not successful, although aldononitrile, trifluoroacetate, and trimethylsilyl derivatives were used by some investigators (Fox and Black 1994).

5.4.2.1.3 Proteins

Proteins are the functional units of all biological organisms, and protein signatures represent an alternative to nucleic acid-based techniques for bacterial agent identification. Standard slab gel techniques are well established in the microbiology and biochemistry arenas for analysis of proteins, and can show differences in the proteome or protein separation patterns that should help characterize the organisms. Nevertheless, these methods are time and labor-intensive. Therefore, faster approaches are widely investigated.

In general, proteins isolated from lysed bacterial cells will contain other constituents such as lipids, nucleic acids, and polysaccharides as contaminants. Because proteins are often insoluble in their native state, breaking interactions involved in protein aggregation, e.g., disulfide/hydrogen bonds, van der Waals forces, ionic and hydrophobic interactions, enable disruption of proteins into a solution of individual polypeptides and thus promote their solubilization (Wilkins et al. 1997). Unfortunately, the presence of buffers, chaotropes, detergents, or cocktails of proteinase inhibitors that are usually added to aid in protein extraction and to preserve the integrity of a proteome may interfere with further processing and analysis of proteins. Therefore, they have to be removed from the sample before sample introduction into MS. Due to the relatively low molecular mass of these additives and many other cellular contaminants in comparison to M_r of proteins, size-exclusion approaches are frequently used to remove them from protein samples. These methods include size-exclusion chromatography (SEC), dialysis, or ultrafiltration. In addition, both ion-exchange and reversed-phase chromatography are frequently used to clean samples before MS analysis and there are many commercially available solid-phase microextraction systems, which enable simple and rapid extraction and purification of proteins (e.g., ZipTip, ZipPlate, Gelloader, or MassPREP PROtarget) (Wallman et al. 2004).

Commercially available capillary chromatography systems for proteomics applications usually include trapping precolumns, where the sample is purified, desalted, and pre-concentrated prior to injection onto a capillary column. In general, high-performance liquid chromatography (HPLC) is an important separation technique for analysis of proteins and peptides because it can easily be coupled to MS.

Moreover, the compatibility of solvents used in the reversed-phase chromatographic separations with ESI makes this hyphenated technique most commonly used in the final stage of proteomics analysis.

A gel-free analysis of extracted proteins is usually performed using bottom-up or top-down MS-based proteomics approaches. For bottom-up type of analysis, a mixture of proteins is digested into peptides, which are next separated, ionized, and mass analyzed using MS instruments to obtain sequence information. The top-down approach relies on separation of extracted proteins and the critical component is the measurement of their molecular masses and partial sequences.

Other LC subtypes, including size-exclusion, ion-exchange, and affinity separations, are commonly used during consecutive steps of sample preparation, clean-up, enrichment, and prefractionation. Most chromatographic approaches are tolerant to moderate concentration of contaminants, such as weak buffers. In this part, we will summarize several exemplary sample pretreatment approaches used prior to injection onto LC column. First, it should be noted that samples injected onto chromatographic column cannot contain insoluble particles or dispersed molecules that may cause column clogging and malfunction. Such contaminants are usually removed by centrifugation and/or sample filtration using spin filters. In addition, samples should not contain buffers affecting LC separation, e.g., samples injected onto a column should not be dissolved in buffer with higher eluting strength than that of mobile phase. High concentration of detergents should be avoided in the case of RP separation whereas samples injected on the ion-exchange column should not contain high contraction of background salts and other ionic contaminants that might disturb ionic equilibrium. Volatile buffers such as ammonium acetate or ammonium bicarbonate (ABC) are recommended in this case.

5.4.2.2 Autonomous Microfluidic Sample Preparation System

Recently, Stachowiak et al. (2007) described an autonomous miocrofluidic sample preparation system for protein profile-based detection of microbial cells and spores. This system was designed to prepare aqueous samples containing bacterial spores and cells in dilute concentration for proteomic signature detection and identification using a capillary gel electrophoresis instrument. According to the authors, the samples used to test the system were intended to simulate those produced by a conventional wetted cyclone aerosol collector such as the SASS 2000+ (Research International, Monroe, WA). The sample preparation system accepted relatively large volume (2 mL) aqueous samples containing dilute, intact microorganisms and delivered a small volume (5 μL) of concentrated, solubilized proteins that were additionally fluorescently labeled due to requirements of the detection system used. These proteins represent proteomes of the collected organisms and could be further analyzed using MS-based methods.

Generally, the sample-processing system consists of a series of microfluidic modules that perform specific consecutive processes on the collected aqueous samples. Individual modules were connected by a fused-silica capillary, while miniature, motorized pumps and valves controlled by custom software written using LabVIEW (National Instruments, Austin, TX) accomplished fluid metering. A schematic representation of the system is shown in Figure 5.5. In this graph, by following the blue

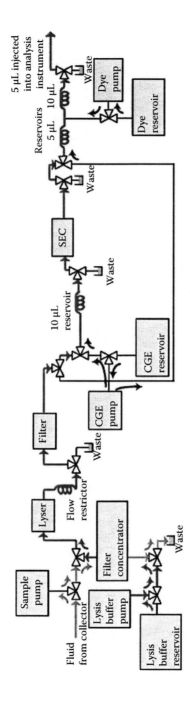

FIGURE 5.5 Schematic representation of the ASP system. Fluidic connections between ports perpendicular to each other are made and broken as the valve actuates. Ports parallel to one another are never connected. (From Stachowiak, J.C. et al., *Anal. Chem.*, 79(15), 5763, 2007.)

path, one can trace the processing of a dilute sample that entered the system from the aerosol collector and was concentrated on a filter. Following the red path, the concentrated sample was flushed from the filter by lysis buffer from the lysis buffer pump, which pushed the sample through a lyser, flow restrictor, filter, and into a 10 μL reservoir. Flow was redirected, and the sample within the reservoir was pushed through a SEC module by a pump containing sodium borate buffer with a detergent (CGE). Inside the chromatography module, the lysis chemicals were separated from the solubilized proteins in the sample. Elution from the module was timed so that once the protein-containing fraction eluted from the module into a 5 μL reservoir, the flow from the CGE buffer pump was diverted to bypass the SEC module, leaving the lysis chemicals within the lysis module. Pushed by the CGE buffer pump, the sample passed a tee, where it was mixed with fluorescamine dye. By avoiding the latter step, proteins could be processed for MS analysis, e.g., by simply replacing dye reservoir with a trypsin solution or by using an in-line trypsin reactor. During sample analysis, the system components were flushed with CGE buffer from the CGE pump, preparing the ASP system for the next run. Figure 5.6 is a photograph of many of the components of the ASP system, such as pumps, valves, reservoirs, a lyser, and an SEC module (Stachowiak et al. 2007). The cycle time for the sample preparation was ~5 min.

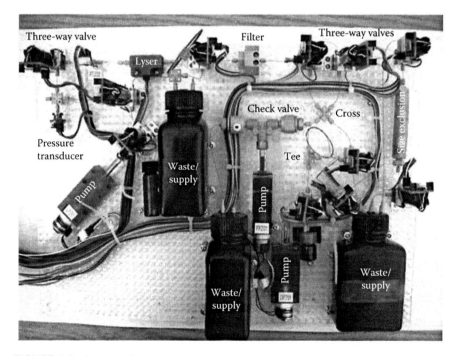

FIGURE 5.6 Integrated components of ASP system. They include motorized pumps and three-way valves, pressure transducers, a thermochemical lyser, filters, a SEC module, check valves, and microfluidic interconnects such as crosses and tees. (From Stachowiak, J.C. et al., *Anal. Chem.*, 79(15), 5763, 2007.)

5.4.2.2.1   Nucleic Acids

Nucleic acids, which reside within a cell or spore, are released by cell/spore rupture and must be extracted from cellular debris and environmental contaminants that may potentially interfere or inhibit amplification procedures.

5.4.2.2.2   Nucleic Acid Extraction

The efficiency of extraction and purification of DNA or RNA from cells or spores of potential biothreat agents influences the sensitivity, reproducibility, and accuracy of any nucleic acid detection method. Substances in complex sample matrices can inhibit hybridization and enzymatic reactions, degrade the nucleic acid, and reduce the efficiency of cell or spore lysis. In large samples, the nucleic acid must also be concentrated, often over 1000-fold, into a smaller volume appropriate for nucleic acid analysis. There are many kits commercially available for purifying nucleic acid (Chua 2004). The following are a few examples of rapid nucleic acid purification methods that have been developed to purify and concentrate nucleic acid from complex sample matrices for MS analysis.

A number of effective methods have been developed over the years to lyse cells and remove contaminants from samples. Cell lysis using chaotropic salts (e.g., guanidinium isothiocyanate) is one effective method and is often followed by capture of the DNA (and not the contaminants) on silica-based substrate under high-salt conditions. Once the contaminants are washed away, the DNA is then eluted from the silica in a low-salt buffer and then analyzed. This and similar methods are available commercially, and they are often performed in plastic tubes using a centrifuge or vacuum or pressure to transport and process the samples. These commercial techniques, while effective, require between 30 min and 2 h and are therefore not suitable for the detect-to-warn application.

The Cepheid GeneXpert and Biothreat agent detection system (Cepheid, Sunnyvale, CA) integrates sample preparation, PCR, and detection into a disposable cartridge. An instrument automatically processes the cartridge, allowing sample preparation in fewer than 5 min and detection (four-color real-time PCR) in 25 min. No special skills are required to use the system. Large volumes (100 μL to 5 mL) of raw sample can be handled, and up to four targets can be detected simultaneously per cartridge. The system automatically sonicates and purifies the sample; PCR reagent is then added to the extracted nucleic acid, and the mixture is dispensed into a PCR tube (McMillan 2002). Swab extracts, body fluids, and *B. anthracis* (Ames strain) spores have also been detected using the system.

Extraction of many types of samples for nucleic acid purification has also been automated using the MagNaPure compact or LC instrument (Roche Applied Science, Indianapolis, IN). Eight or 32 samples, respectively, can be purified in 1–3 h. The purification process involves the binding of nucleic acid to surfaces of magnetic glass particles, extensive washing, and then elution of nucleic acid. The MagNaPure instrument has been used successfully to purify nucleic acid from clinical samples, ticks (Exner and Lewinski 2003), and powders (Luna et al. 2003).

Methods for using electric fields to attract, focus, or separate DNA, bacteria, and other particles in solution have been demonstrated and may work within time scales that are suitable for mobile lab applications. The majority of biological particles are amphoteric; that is, they can be net negative, net positive, or net neutral, depending on the local pH. The pH at which a particle is neutral is called the isoelectric point (pI). At this pH, the particle experiences no net force when an electric field is applied. By generating a suitable pH gradient parallel to an electric field, particles in the field will migrate to their pI and remain fixed at that position. Finally, biological particles also become polarized when placed in an electric field. These induced dipoles can lead to net migration in an applied, nonuniform field, which is referred to as dielectrophoresis (Schnelle et al. 2000). The net velocity of a particle within such a nonuniform field depends on its dielectric function relative to that of the surrounding media as well as on the frequency and strength of the applied electric field. In batch mode, cells can travel at velocities up to 0.1 mm/s.

ESI-FTICR-MS has been used to analyze two types of genetic markers, that is, alleles from variable number tandem repeats (VNTRs) and single-nucleotide polymorphisms (SNPs) (Van Ert et al. 2004). To achieve this goal, DNA was amplified and fragment size was kept to less than 200 base pairs (bp) to stay within the optimum size range for the MS procedure. Analysis of VNTR and SNPs PCR products from *B. anthracis* using EIS-FTICR-MS produced results comparable to traditional gel electrophoresis.

5.4.2.2.3   DNA Isolation from Aerosol Samples

To obtain MS-derived base composition signatures of microbial genomes, Hofstadler et al. (2005) used air samples collected directly into a buffer/detergent liquid matrix or onto polyester fiber filters, which were immersed in phosphate-buffered saline (PBS) containing 0.1% Tween-20 detergent. The filters and solution were shaken by hand for 30 s and the resulting solution was filtered through a 0.2 μm filter. The resulting filters were then subjected to bead beating by placing the filter in a 1.5 mL tube containing 100 μg of 0.7 mm zirconium beads and 350 μL of a lysis buffer. The beads were shaken on a mixer mill for 10 min and then spun briefly to settle the beads and larger particles. The supernatant, containing the nucleic acid material, was then used as the starting material in a Qiagen DNeasy Tissue Kit isolation on a Qiagen BioRobot 8000 (Qiagen, Valencia, CA) following the manufacturer's protocol. The extracted DNA is first amplified with broad-range primers and PCR-derived amplicons are submitted for MS analysis (Figure 5.7).

5.4.2.2.4   Preparation of 16S rRNA Cleavage Products for MS Analysis

Genomic DNA was released by boiling lysis of bacterial cells, and ca. 25 ng total genomic DNA of such lysate was used for subsequent PCR using primers for the 16S rDNA gene. The conventional DNA isolation and PCR steps required ~2 h, while subsequent transcription and enzymatic cleavage, less than 30 min. All steps are amenable to high-throughput implementation. To limit the complexity of MS spectra acquired, only the primer pairs for amplification of ca. 400 and ca. 500 bp of 16S rDNA from most organisms were used. All forward primers employed also contained an extension for incorporation of a T7 RNA polymerase promoter sequence, and all reverse primers were "5'-tailed" with the reverse complement of a sequence used for single point internal mass calibration. Conventional PCR thermal cycling conditions

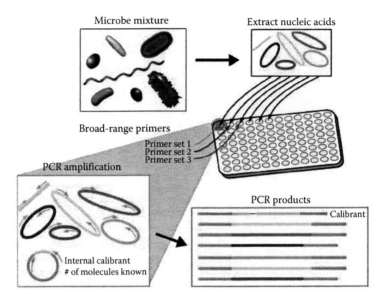

FIGURE 5.7   TIGER concept of operation. Nucleic acid extracts from the sample of interest (e.g., air sample, clinical specimen, food product) are amplified with broad-range PCR primers, and PCR-derived amplicons are analyzed using MS. (From Ecker, D.J. et al., *J. Assoc. Lab. Automation*, 11(6), 341, 2006.)

were used for ~105 min. Following PCR, reaction mixtures were treated with DNA exonuclease I to digest any unincorporated single-stranded primers. The resulting mixture was then used directly as template for in vitro transcription with T7 RNA polymerase. Following transcription, ribonuclease T1 was added and transcripts were cleaved (after G residues) for 5 min at 37°C. Finally, RNA cleavage product mixtures were desalted by reverse phase purification. In the final step, RNA cleavage products were eluted with a suitable MALDI matrix (Jackson et al. 2007).

5.4.2.2.5   RNA Extraction, Reverse Transcription, and PCR of Viral Nucleic Acids

RNA from samples containing RNA viruses may be isolated using commercially available kits like Trizol. For example, Sampath et al. (2005) performed reverse transcription of the purified RNA with random primers and RNase inhibitors and products of the reaction were used for the PCR reactions. PCR reactions were performed in microtiter plates and thermocyclers. The standard PCR conditions were used to amplify viral sequences, and the amplified products were desalted before analysis by ESI-MS (Figure 5.7).

5.5   SAMPLE INTRODUCTION FOR MS ANALYSIS

Advances in sample preparation for ESI include the use of chips and microfluidic devices which may integrate the successive preparation steps. For instance, in the case of protein analysis, digestion followed by separation of peptides by 1D-, 2D-LC, or capillary electrophoresis are used as sample introduction devices, which are directly coupled to a nano-ESI tandem MS instrument (Ramsey and Ramsey 1997). Small-volume samples may be handled and sensitive analyses are possible. Peterson et al. (2003) have introduced an integrated device which has a 40 nL microcolumn with immobilized trypsin for protein digestion and a SPE microcartridge for desalting/concentration of digested peptides. Volumes in the nanoliter range are manipulated, and digestions are produced in 1 min.

5.5.1 ON-CHIP SAMPLE PREPARATION, CONCENTRATION, AND FRACTIONATION

Recently, a microfluidic chip for peptide analysis has been introduced on the market by Agilent Laboratories in Palo Alto, CA (Yin et al. 2005), that simplifies the nano-LC-MS system. In this approach, the traditional use of an enrichment column, a separation column, a nanospray tip, and the fittings needed to connect these parts together has been replaced by a microfabricated device. A single microfluidic chip integrates these components, thus eliminating the need for conventional LC connections. The chip was fabricated by laminating polyimide films with laser-ablated channels, ports, and frit structures. The enrichment and separation columns were packed using conventional reversed-phase chromatography particles. A face-seal rotary valve provided a means for switching between sample loading and separation configurations with minimum dead and delay volumes while allowing high-pressure operation. The LC chip and valve assembly were mounted within a custom electrospray source on an ion trap MS. The overall system performance was demonstrated through reversed-phase gradient separations of tryptic protein digests at flow rates between 100 and 400 nL/min. Microfluidic integration of the nano-LC components enabled separations with sub-femtomole detection sensitivity, minimal carryover, and robust and stable electrospray throughout the LC solvent gradient.

MALDI-MS is a widespread and powerful analytical tool in the analysis of peptides, proteins, and other biomolecules like oligonucleotides, carbohydrates, natural products, and lipids. UV-MALDI-MS provides superior sensitivity, enabling detection of femtomole–attomole concentrations and significant tolerance against contaminations, such as salts and/or common buffers; therefore, samples are deposited directly on the target manually or by using robotic devices.

There have been many attempts to develop a biosensor based on MS. Most of these approaches utilize MALDI-TOF. These systems are reagentless and rapid, require only a small sample volume, and potentially require little sample preparation. MALDI-TOF-MS is theoretically capable of identifying all types of biological agents, including viruses, bacteria, fungi, and spores. Drawbacks to this approach include the requirement for a highly concentrated sample (10^5–10^7 cells/mL for whole-cell analysis), the need to develop complex spectral fingerprints for every target agent, and the possible lack of specificity in complex matrices or mixtures of targets. Therefore, any sample processing aimed at removal of contaminants and fractionate components is needed. Among them, a HF FFF is being investigated as one way of increasing specificity by separating particles in a complex matrix prior to analysis by MALDI-TOF-MS (Lee and Williams 2003; Reschiglian et al. 2004).

5.5.2 MICROFLUIDIC DEVICES

There is wide spread interest in coupling microfluidic devices to ESI or MALDI-MS. Indeed, if a fully integrated method, comprising chemical processing, sample pre-concentration and clean-up, and 2D separations could be integrated with MS detection, the result could revolutionize the field of proteomics. This revolution is on hold; however, as current microfluidic-MS interfaces have not yet been widely adopted they continue to be plagued with technical challenges. Here, we survey the different kinds of interfaces reported in the literature, and discuss the most promising geometries and applications going forward. We contend that nanoelectrospray ionization is the most likely candidate for a robust interface between microfluidics and MS. This assertion springs from the obvious similarities between the conventional technique of interfacing HPLC eluent to a spectrometer by means of pulled glass nanospray tips, and the linear geometry of microfluidic channels. A variety of strategies for fabricating such devices have been reported, in which proteomic sample solutions are pumped through microchannels pneumatically or by electroosmotic flow (EOF) at ca. 100–300 nL/min. Samples are typically dissolved in low pH buffers modified with organic solvents suitable for positive mode MS, with detection limits in the femtomole–attomole range. These methods can be broadly classified by how the electrospray is generated, including (1) direct spray from channels (Ramsey and Ramsey 1997; Wang et al. 2004); (2) spray from mated, conventional tips (Ssenyange et al. 2004); and (3) spray from microfabricated tips (Licklider et al. 2000; Xie et al. 2005; Yin et al. 2005). Electrospray directly from a channel (i.e., the unmodified edge of a device) is the easiest approach for interfacing with MS. Despite some advances, spraying directly from the edges of chips has been largely abandoned due to reduced sensitivity and decreased resolution when coupled to separations.

5.5.3 LC-MALDI

The fully automated collection of LC separations and automated acquisition of the MALDI-MS spectra is a quite recent development. Such combination of LC with MALDI-MS has several advantages over the more widespread online LC-ESI-MS approach, including the ability to store an LC-based separation on a MALDI target plate so that MS analyses can be conducted without time constraints and samples can be archived. New types of MS instruments, especially the new MALDI-TOF-TOF instruments, have really boosted interest in LC-MALDI interfacing. The robotic interfaces for LC-MALDI-MS promise to provide higher throughput, capacity, and speed in identifying molecules. Table 5.1 lists some commercially available instruments that interface LC and MALDI-MS. Robotic interfaces collect eluents directly from LC columns as the fractions appear and transfer them onto a MALDI target plate.

LC fractions can be deposited onto a MALDI target plate as discrete spots or continuously forming a "snail trail." In noncontact deposition (e.g., Agilent and Shimadzu systems) the bottom of the droplet touches the target plate, while in the contact method (Gilson) the needle touches the surface to deposit the droplets. In addition, all the commercial instruments offer online matrix addition.

TABLE 5.1
Commercially Available Instruments for Interfacing LC and MALDI-MS

Product	1100 series Microcollector/spotter	Microfraction collector	223 MALDI FC/spotter	Proteineer FC	AccuSpot
Company	Agilent Technologies www.agilent.com	LC Packings-Dionex Co. www.lcpackings.com	Glison Inc. www.gilson.com	Bruker Daltonics Inc. www.bdal.com	Shimadzu Biotech USA www.shimadzu.biotech.net
Volume capacity	>100 nL	>5 nL	0.020–5000 μL	nL to μL	0.1–0.5 μL/min
Deposition mode	Contact mode	Spotting, streaking	Near-contact to contact dispensing	Sample/matrix solution spotted on target plate	Half-contact (patented plate sensor technology)
Sample capacity	496-well plates or four MALDI targets	Up to 3456 spots on six high-density targets	12 Applied Biosystem MALDI plates	4×384 spot MALDI target	9×384 sample MALDI plates
Deposition speed (time/sample)	3 s	Down to 5 s	1.2 s	Flow and peak width dependent	5–60 s

The success of analyte transfer onto a MALDI target depends on the flow from the LC column and the spotting frequency onto the target. If the peaks on the chromatogram are very sharp, spotting frequencies onto the MALDI target may only need to be a few seconds per spot. However, analytes from the LC column can collect in the capillary and cause carryover from spot to spot, and the resolution from the chromatographic separation can be lost. The droplets can be made so tiny that they are held together by surface tension, or the plate can be treated to be largely hydrophobic, with small, confined areas that are hydrophilic. Once analytes are deposited from the LC onto the MALDI target plate, the plate can be considered a "readdressable record" in which analyses can be carried out several times over an extended period of time. But some experts worry about the quality of the data in subsequent analyses because analyte degradation may occur.

In order for LC-MALDI-MS to produce useful, timely data on large quantities of molecules, its throughput must be further increased by using lasers firing with high frequency (e.g., 2 kHz) and by increasing a data acquisition speed of MALDI instruments. Recently, a high-throughput online capillary array-based 2D-LC system coupled with a MALDI-TOF-TOF-MSP analyzer for comprehensive proteomic analyses has been developed by Gu et al. (2006). In this design, one capillary SCX chromatographic column was used as the first separation dimension and 18 parallel capillary RPLC columns were integrated as the second separation dimension. Peptides bound to the SCX phase were "stepped" off using multiple salt pulses followed by sequential loading of each subset of peptides onto the corresponding precolumns. After salt fractionation, by directing identically split solvent-gradient flows into 18 channels, peptide fractions were concurrently back-flushed from the precolumns and separated simultaneously with 18 capillary RP columns. LC effluents were directly deposited onto the MALDI target plates through an array of capillary tips at a 15 s interval, and then CHCA matrix solution was added to each sample spot for subsequent MALDI experiments. This new system allows an 18-fold increase in throughput compared with a serial-based 2D-LC system. The high efficiency of the overall system was demonstrated by the analysis of a tryptic digest of proteins extracted from liver tissue. A total of 462 proteins were identified, which proved the system's potential for high-throughput analysis and application in proteomics investigations (Gu et al. 2006).

A very promising approach for interfacing LC with AP-MALDI-MS was recently reported by researchers from the Swiss Federal Institute of Technology in Switzerland and MassTech Inc. in Columbia, MD (Daniel et al. 2005). Because the operation of the AP-MALDI eliminates the need for introducing the sample into the vacuum, this permits the use of volatile matrices like water and opens the possibility for online interfacing LC and MS. They presented two different strategies: (a) the first method is a flow injection liquid AP-UV-MALDI with a laser beam focused onto droplets forming at the end of exit capillary (Figure 5.8) and (b) the second strategy is based on a single-droplet AP-IR-MALDI (Figure 5.9). Compared with previous similar research, the detection limit was improved 10 times to 8.3 fmol using a solution of 50 nM peptide with 25 mM CHCA and the applicability of this method to measure oligosaccharides, peptides, and proteins was demonstrated. The overall sensitivity achieved using an IR laser with a repetition rate of 5 Hz was low. Nevertheless, this

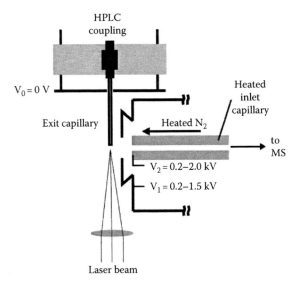

FIGURE 5.8 Flow injection setup used for AP-UV-MALDI. The analyte/matrix mixture is fed into the fused-silica exit capillary by a HPLC coupling. A pulsed laser beam is focused onto the droplets forming at the end of the exit capillary, desorbing, and ionizing the matrix and analyte. The ions produced are guided by electric fields to the inlet of the MS interface. (From Daniel, J.M. et al., *Anal. Bioanal. Chem.*, 383(6), 895, 2005.)

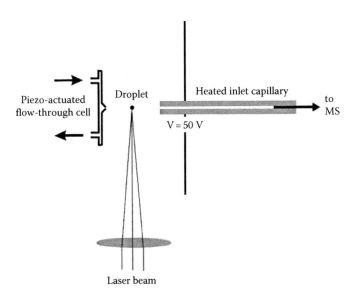

FIGURE 5.9 Setup used for single-droplet IR experiments. The droplets are generated by a piezo-actuated flow-through cell. The laser, synchronized to the droplet ejection, is focused onto the flying droplet desorbing the analytes. (From Daniel, J.M. et al., *Anal. Bioanal. Chem.*, 383(6), 895, 2005.)

method shows great potential for coupling LC to MS if a higher repetition rate IR laser is used, because there is no additional mixing step needed prior to the desorption and standard water-based liquid phases could be used.

5.6 ONE-STEP APPROACH FOR PROCESSING AND ANALYZING PROTEINS

A "one-step" method for processing proteins and isolating peptide mixtures from bacterial samples is presented for LC-MS/MS analysis and data reduction. The conventional in-solution digestion of the protein contents of bacteria is compared to a small disposable filter unit placed inside a centrifuge vial for processing and digestion of bacterial proteins. Each processing stage allows filtration of excess reactants and unwanted by-products while retaining the proteins. Upon addition of trypsin, the peptide mixture solution is passed through the filter while retaining the trypsin enzyme. The peptide mixture is then analyzed by LC-MS/MS and the ABOid™ algorithm for a comparison of the experimental unique peptides to a constructed proteome database of bacterial genus, species, and strain entries. The concentration of bacteria was varied from 10×10^7 to 3.3×10^3 cfu/mL for analysis of the effect of concentration on the ability of the sample processing, LC-MS/MS, and data analysis methods to identify bacteria. The protein processing method and dilution procedure result in reliable identification of pure suspensions and mixtures at high and low bacterial concentrations.

Effective sample processing is a critical element for a program that requires the separation and analysis of complex biochemical components from their intact biological host such as bacteria. When only a subset of biochemical entities defines a sample interrogation and interpretation methodology, the challenge lies in producing conditions that yield their greatest efficiency of capture and extraction from the cellular milieu for subsequent delivery into the detector. Efficiency aspects can include the degree and amount of biomolecule analyte extraction, complexity of the physical processing conditions and modules, ease of biomolecule subset concentration with interference removal, analyte transfer away from the modules, and ion space-charge consideration in the MS ion trap detector. These attributes especially can help in the rapid analysis from first responder, environmental or terrorist, and medical/clinical scenarios where efficient sample handling and sensitivity are of paramount importance in biosafety situations.

Processing of microorganism proteins for proteomics analysis over the past decade has developed into three main methods. Initially, all processing methods rely on cellular lysis to provide access to the vast numbers of proteins. Then either online or off-line transfer of the protein-laden supernatant is performed for additional processing. An important method of protein separation that has found extensive utility is 2D polyacrylamide gel electrophoresis (Quadroni and James 1999; Lopez 2000; Lambert et al. 2005).

However, processing of the many separated proteins is performed by excising a protein spot from the gel with subsequent purification, concentration, and trypsin digestion. The physical manipulations of the PAGE method are very time-consuming, and there are inherent limitations such as not providing for the realization of proteins with low and high molecular weights, low and high p*I* values, and the capture of nonpolar membrane-bound proteins (Washburn et al. 2001; Ihling and Sinz 2005). Alternative methods have been developed for direct protein processing with liquid and/or stationary supports (Washburn et al. 2001; Vollmer et al. 2003; Wu et al. 2003; Dai et al. 2005; Salzano et al. 2007; Malen et al. 2008).

Preprocessing consisted of protein precipitation, denaturation using concentrated urea to remove the protein secondary and tertiary structures, DTT disulfide reduction, and alkylation steps. Trypsin digestion was then performed either in solution or on solid-phase porozyme media. The peptide supernatant was concentrated and/or introduced into a system which consisted of a SCX-RPLC-LC-MS/MS for comprehensive peptide separation and detection. The detailed sequence of steps usually followed different degrees of online status, where some steps were manual and others were performed from one step to the next with sample transfer from one module to the next module.

Craft and Li (2005) integrated a heated clean-up and digestion module for cytochrome *c* and bovine serum albumin protein standards prior to LC-MS/MS analysis. Ma et al. (2009) took this a significant step further and combined thermal denaturation reduction, digestion, and peptide pre-concentration of protein standards and mouse liver protein extract for LC-MS/MS peptide characterization.

In another method, many of the initial procedures were combined into a "one-step" system. Ethier et al. (2006) constructed an SCX proteomic reactor to accept the cell lysate. Sequential processing steps took place; however, it was unclear as to the fate of the residual reactants and by-products in the denaturation, DTT reduction, iodoacetamide (IA) alkylation, and trypsin digestion steps. Evaporation and concentration completed the peptide processing, and the peptides were presented to an LC-MS/MS for separation and analysis. The concern here is that the reagents for each step prior to the LC-MS/MS were not removed or separated, and this was also of concern to Ma et al. (2009). It was possible that the reduction/alkylation/trypsin digestion reactants and residual products remained in the reactant SCX column to cause potential suboptimal conditions for protein processing.

This approach depends on proteome database analyses between bacterial strains and their multiple respective strains resident in a bacterial proteome database. MALDI-MS literature also provides ample evidence of proteome studies on bacterial strains for identification purposes. *S. aureus* (Walker et al. 2002), *E. coli* (Bright et al. 2002), Campylobacter (Mandrell et al. 2005), *Bacillus* species (Leenders et al. 1999; Ryzhov et al. 2000; Dickinson et al. 2004) including *B. anthracis* (Castanha et al. 2006), and *Rhodococcus erythropolis* (Teramoto et al. 2009) were investigated for strain differentiation by MALDI-MS. Sample preparation and handling procedures essentially relied on either intact, lysed, or sonicated cells and subsequent mixing with a suitable organic matrix.

The purpose of this section was to seek procedures toward a comprehensive and convenient processing of bacterial cell lysate proteins into a "one-system" design with no off-line components including the removal of reactants and by-products. FA was purchased from Burdick and Jackson Laboratories, Inc. (Muskegon, MI). Sequencing grade-modified trypsin was purchased from Promega (Madison, WI), and myoglobin (from equine heart) was purchased from Sigma-Aldrich (St. Louis, MO).

5.6.1 MICROBIOLOGICAL SAMPLE PREPARATIONS

B. subtilis 168, *S. aureus* ATCC 12600, and *E. coli* K-12 preparation and handling procedures can be found elsewhere (Jabbour et al. 2010a,b).

5.6.2 BACTERIAL DILUTION

The ability of the peptide analysis sequencing and data reduction methods to effect bacterial differentiation was tested by varying the concentration of bacteria from 10×10^7 to 10×10^4. The dilution protocol was a standard serial dilution from different vials. Two milliliters of a 2×10^7 cfu/mL stock suspension of bacteria was placed in a separate vial. This 2 mL volume was sonicated for a total of 3 min (Branson Digital Sonifier, Danbury, CT 06810) consisting of 20 s pulse-on and 5 s pulse-off intervals. A 25% amplitude was chosen in order to lyse the cells. To verify that the cells were disrupted, a small portion of the lysate was reserved for 1D gel analysis. Five hundred microliters of the sonicated suspension was transferred to another vial, and 500 μL of ABC buffer was added to yield 1 mL of an equivalent 10×10^7 cfu/mL. One hundred microliters from the sonicated 10×10^7 bacterial suspension was transferred into another vial, and 900 μL of ABC buffer was added to yield 1 mL of an equivalent 10×10^6 cfu/mL suspension of bacteria. Two more iterations were performed to produce 10×10^5 and 10×10^4 cfu/mL equivalent suspensions of bacteria. Note that all four dilutions resulted from one stock sonicated bacterial suspension.

For all four dilutions, each was performed three times to yield a standard deviation for each replicate series. The dilution method was investigated for its impact on the ability to define the sample bacterium from a proteomics statistical analysis.

5.6.3 BACTERIAL MIXTURE ANALYSIS

Experiments have been conducted regarding the undigested protein and bacterial debris after each step (Jabbour et al. 2007, 2008). These compounds and components were directed away and where equal volumes of 10×10^7 cfu/mL suspensions of *E. coli* upstream from the analytical peptide separation and detection LC-MS/MS system. A 3 kDa molecular weight cutoff (MWCO) membrane was used without LC column or separation components; therefore, the proteins were retained during processing. Once peptides were generated on the 3 kDa MWCO membrane by trypsin digestion, they were passed through the membrane and loaded onto the analytical LC column for mass spectral analysis. The trypsin enzyme was retained by the membrane. In a similar methodology concept, Wisniewski et al. (2009a) outlined a filter-aided sample preparation (FASP) method detailing a "one-pot" procedure that highlighted the removal of low-molecular-weight reagents and impurities while retaining the high-molecular-weight species. The retained protein extract was then digested so as to pass through the peptides and retain the trypsin species. Wisniewski et al. (2009a) applied their FASP method to bovine serum albumin and HeLa cells. We extend the FASP work to the isolation of a bacterial protein extract and its peptide mixture and that of a mixture of bacteria. In this manner, the "one-step" analysis further compared with the conventional in-solution digestion method for identification and reproducibility in a range of concentrations for pure bacteria and a mixture of bacteria.

5.6.4 Methods for One-Step Testing

5.6.4.1 Materials and Reagents

ABC, DTT, urea, ACN HPLC grade, K-12, *S. aureus* 12600, and *B. subtilis* 168 were added into a vial to result in a 3.3×10^6 cfu/mL for each bacterium. This experiment was conducted in a double-blind situation (Jabbour et al. 2010b). The mixture was sonicated for 3 min to release the cellular proteins for further processing. The dilution series protocol was conducted to test the ability of the procedures and data analysis methods to detect and define the number and identity of organisms present in the mixture at different concentrations.

5.6.4.2 Bacterial Processing

Two methods were used to isolate and digest the protein portion of the bacteria. The first method was basically the in-solution digestion technique 13 with some modifications. One milliliter of a given dilution and lysed bacterial suspension was placed into vial 1 (V1). V1 was centrifuged for 10 min at 14,000 rpm. The protein-laden supernatant was pipetted into a Microcon YM-10 filter unit (Millipore #42406) with a 10,000 Da molecular weight cutoff, and the filter unit was inserted into V2. Vial V2 was centrifuged to remove the liquid. The proteins were retained on the filter surface and the filtrate was discarded. Fifty microliters of 100 mM ABC buffer was added to dissolve the proteins, and the liquid was retained on the filter. The filter unit was inserted into V2 and vortexed. The filter unit was then turned upside down in V3 and the assembly was centrifuged to deposit the ~50 μL protein-laden liquid onto the bottom of V3. Thirty microliters of DTT (10 μg/mL) and 270 μL of 8 M urea were added into V3 and placed in a 45°C oven for 1 h to denature the protein. One milliliter of ABC buffer was added to V3 and evaporated in a SpeedVac. Then 5 μL of trypsin (1 mg/mL), 20 μL of ABC buffer, and 5 μL of ACN were added and vortexed briefly to dissolve the protein on the inside walls of V3. Protein digestion occurred overnight at 37°C on an orbital shaker at 55 rpm. Sixty microliters of 5% ACN/0.5% FA was added with 2 min of vortexing for sample mixing to quench the digestion. The liquid was evaporated in a SpeedVac. The peptides were reconstituted with 100 μL of LC buffer (vide infra). Ten microliters of the peptide solution was injected into the LC column.

The solution-digestion filter method was the alternative set of procedures used for bacterial protein isolation and digestion (Jabbour et al. 2007, 2008). This method begins with the in-solution digestion method and instead of turning the filter upside down, the filter remained right side up in V2. Fifty microliters of ABC buffer was added onto the filter and centrifuged to effect a buffer exchange. The proteins were retained on the filter. DTT and urea were added onto the filter, and the V2-filter assembly was placed in a 45°C oven for 1 h. One hundred microliters of ABC buffer was added to dilute the urea, and V2 was centrifuged. Another 100 μL of ABC buffer application was performed, and V2 was centrifuged. Trypsin was then added with the ABC/ACN solution as in method A. Protein digestion occurred overnight in a 37°C oven on an orbital shaker at 55 rpm. Sixty microliters of 5% ACN/0.5% FA was added with 2 min of vortexing to quench the digestion. V2 was replaced with V3, and the unit was centrifuged to collect the peptide supernatant. Ten microliters of the supernatant was injected into the LC column. LC-MS/MS Analysis and Protein Database procedures are outlined elsewhere (Jabbour et al. 2007, 2008).

5.6.5 "One-Step" Discussion

An objective of this work was to reduce the number of surfaces and vials necessary to process bacteria into an isolated protein extract and peptide mixture when compared to the standard in-solution digestion technique. This modified, somewhat streamlined procedure is labeled the solution-digestion filter method. A proteome-based identification of the mass spectral peptide response would provide an adequate measure of the efficacy of the solution-digestion filter method.

Different concentrations of a protein mixture extract from the same bacterium can yield different qualitative and quantitative information. This can affect performance of the proteome data analysis algorithms, and as such, bacterial concentrations ranged from 10×10^7 to 10×10^4 cfu/mL.

5.7 MYOGLOBIN PEPTIDE RECOVERY EFFICIENCY

Pure myoglobin protein was subjected to two different processing methods. Figure 5.10a presents an LC chromatogram of the myoglobin peptide mixture after processing with the standard in-solution digestion method. Note in particular the circled, broad peak around 90 min. The separation procedure shows that the reagents and buffer components are subjected to lyophilization in the processing steps. However, a portion of the reagents such as DTT, urea, FA, ACN buffer, and undigested myoglobin may be present in a residual capacity in the peptide-laden sample injected into the LC system. This appears to affect the qualitative and quantitative recovery of the peptides as shown in Figure 5.10b, and each procedure was performed in three replicate measurements. Figure 5.10b presents the LC chromatogram of the myoglobin peptide mixture from the solution-digestion filter method. The ordinate scale was normalized for comparison purposes in Figure 5.10a and b. Over twice as much peptide recovery was realized with respect to relative abundance (ordinate-scale information) when the residual reagents and by-products were eliminated. These materials are not observed about the 90 min time frame in Figure 5.10b. Also, a peak at 68–69 min is prominently observed in Figure 5.10b while it is virtually absent in Figure 5.10a. Generally, more peptides are retained with the filter method compared to the in-solution digestion method. This may be partially due to fewer manipulations associated with the procedures in the filter processing steps.

There are three myoglobin misclassified peptides in the filter processing method, and one of them (K.YKELGFQG) is observed from the in-solution digestion procedure. The two additional misclassified peptides are not functions of the procedural steps. Rather, they are products of the trypsin digestion which is a common component for both processing methods. Therefore, this is evidence for the in-solution digestion

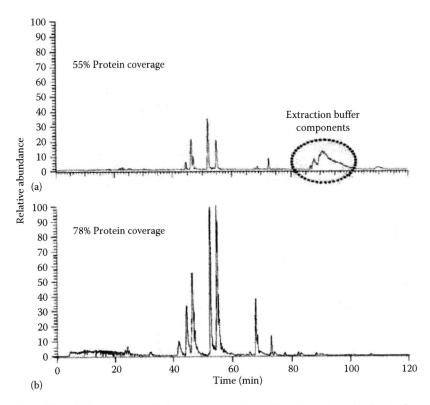

FIGURE 5.10   LC mass spectral chromatograms of peptides from a trypsin digest of myoglobin using the (a) standard in-solution digestion and (b) solution-digestion filter methods. The intensities are scaled relative to the chromatogram in (b). (From Jabbour, R.E. et al., *J. Proteome Res.*, 10, 907, 2011.)

method having a role in prevention, retardation, or inhibition of the admittance into the LC of the K.ALELFRNDI-AAK.Y and K.GHHEAELKPLAQSHATK.H peptides. This interpretation is also likely for three other unique peptides observed from the filter processing method as well as for the two unique peptides observed from the standard in-solution digestion method.

5.8   BACTERIAL PEPTIDE RECOVERY

Only the solution-digestion filter method is utilized for the remainder of the experiments because it outperformed the standard method. Figure 5.11 compares the number of unique peptides using the bacterial dilution protocol with the solution-digestion filter processing method. Separate suspensions of a gram-negative and two gram-positive organisms were used for the dilution comparisons. For each concentration and each organism, three separate experiments were performed. The average is plotted for each bacterial concentration with standard deviation error bars. As predicted, as the amount of lysed bacterial protein preparation (Figure 5.11), the number of unique peptides possessing 95% probability score from PeptideProphet

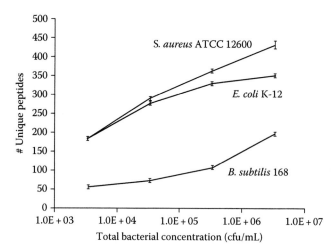

FIGURE 5.11 Number of unique peptides possessing >95% probability score are shown from PeptideProphet using two sample dilution protocols. The solution-digestion filter processing method was used to isolate the peptides for all three bacteria at concentrations of 3.3×10^6, 3.3×10^5, 3.3×10^4, and 3.3×10^3 cfu/mL. The abscissa reflects the total concentration of the experimental bacterial mixture. (From Jabbour, R.E. et al., *J. Proteome Res.*, 10, 907, 2011.)

using two sample dilution protocols increased. The solution-digestion filter processing method was used to isolate the peptides for all three bacteria at concentrations of 3.3×10^6, 3.3×10^5, 3.3×10^4, and 3.3×10^3 cfu/mL. The abscissa reflects that the total concentration of the experimental bacterial mixture decreased, fewer peptides were observed. The concentration of protein in a bacterial cell varies by many decades depending on the specific protein. Proteins with a relatively high concentration usually have a greater probability of detection from their peptide digestion products compared to the proteins with a smaller concentration. However, in an opposite effect, the higher concentration sample will have a greater likelihood for an elution of peptides exhibiting a higher number and amount per unit time through the LC column. In this situation, the relatively lower concentration of certain peptides in a particular LC elution time may be masked by the peptide(s) that exhibits a higher concentration. This can be manifested in the spectra through competition of proton charge. Overall, proton affinities also play a role in the percent of charge acquisition that a particular peptide has with co-eluting peptide(s) at a certain LC column retention time. There is a dilution effect, and the trend is in the same direction for both organisms.

Likewise, the space-charge effects in the limited ion trap volume may also favor more abundant peptides. Ion trap space-charge effects can indiscriminately remove a percentage of peptides eluting from the LC. The higher abundant peptides will most likely be favored in the ion trap fill volume than the lower abundant peptides. This space-charge removal of peptides may cause a greater loss for the less abundant peptides and possibly render less abundant bacterial discriminating peptides undetectable. The trends in Figure 5.11 display an overall decrease in the appearance of peptides with decreasing protein extract concentration, and this may affect the outcome of a detection and identification analysis.

Spores have a greatly reduced need, compared to a typical vegetative bacterial cell, to produce a full complement of proteins in a ready-to-use state for cellular functions. The dormant state of a spore precludes the necessity for the actual presence of many cell function and housekeeping proteins. Thus, it is no surprise to observe the significant decrease in number of peptides for *B. subtilis* spores compared to the peptides for *S. aureus* and *E. coli* in Figure 5.11. It is not common to report this ability for single and mixtures of whole bacteria suspensions at different concentrations.

5.8.1 BACTERIAL MIXTURES AND PEPTIDE RECOVERY

A mixture of three bacteria was prepared with equal volumes of equimolar bacterial stock suspensions. This was a double-blind sample, and we were unaware of the identity and number of organisms in the mixture. We were informed that the total bacterial concentration was approximately 10×10^7 cfu/mL. One milliliter of the 2 mL double-blind mixture suspension was used for identification and dilution studies. Each dilution was separately sonicated and lysed for 3 min.

The histogram in Figure 5.11 presents the results where the peptide mixtures were analyzed and separated into three bacterial species using the ABOid™ program and were further delineated into strains. Plots are represented as the number of unique peptides for each of *E. coli*, *S. aureus*, and *B. subtilis* that were identified and subsequently confirmed following the data reduction procedures for bacterial identification found elsewhere (Dworzanski et al. 2006; Jabbour et al. 2010b). Subsequently, we were told that the double-blind suspension was prepared by adding equal volumes together of equal concentrations of the three organisms. Once the identities and number of bacteria in the mixture were confirmed, this information shaped the standard dilution series in Figure 5.11.

The general trend of a decreasing number of unique peptides, resulting from statistical analysis of the product ion mass spectra, is observed with a decrease in protein extract concentration. An interesting observation is that the peptide trends for all three organisms do not show a "flat-lining" or horizontal trend at the experimental concentrations used. Rather, a decreasing trend is sustained below 10×10^4 cfu/mL. Thus, there are unique peptides even at the relatively low 3.33×10^3 cfu/mL bacterial equivalent protein extract concentration. The correct identity was determined for each of the three bacteria at all four suspension dilutions (Jabbour et al. 2010b).

5.9   BACTERIAL CLASSIFICATION ANALYSIS

Using the solution-digestion filter method, analyses were performed with known and double-blind bacterial suspensions at different concentrations. Lower concentrations of bacteria in general provided a lower amount of peptide recovery. Lower bacterial concentrations of 3.3×10^3 and 3.3×10^4 organisms provided satisfactory identification capabilities. Figure 5.12 provides the dendrogram details of the concentration effect on the qualitative and quantitative aspects of the unique peptide parameters. A level of confidence analysis was performed for the dendrogram results. This process was repeated for different concentrations and replications and results were consistent.

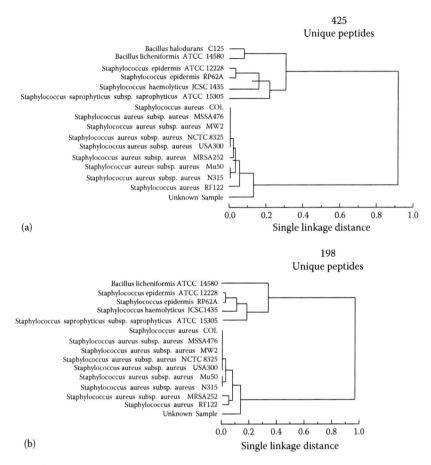

FIGURE 5.12 Dendrograms of double blind suspensions later revealed to be *S. aureus* ATCC 12600. The solution-digestion filter processing and the single linkage Euclidean distance data analysis statistical methods were used. (a) Given concentration of 10×10^7 cfu/mL and (b) 10×10^4 cfu/mL arrived at by serial dilution. (From Jabbour, R.E. et al., *J. Proteome Res.*, 10, 907, 2011.)

5.10 EUCLIDEAN DISTANCE APPROACH

Figure 5.12a provides a classification result from a double-blind study of a dendrogram analysis using the single linkage Euclidean distance method. The concentration of the bacterial suspension was given at 10×10^7 cfu/mL; however, we were not given that information until after our results were reported. Serial dilutions were prepared in decades to a 10×10^4 cfu/mL suspension. The bacterial strain data reduction and analysis resulted in 425 unique peptides for one bacterium. The double-blind sample was closest to pathogenic *S. aureus* RF122 at a relatively high concentration of 10×10^7 cfu/mL. At a lower concentration (Figure 5.12b), pathogenic *S. aureus* MRSA252 and *S. aureus* RF122 were determined to be the closest strains to the sample strain with 198 unique peptides. The bacterium was subsequently revealed to be the pathogenic *S. aureus* ATCC 12600 organism. This particular strain is not contained in the database, because its genome has not been sequenced. The reason that a non-database organism was investigated was to determine if the processing and data reduction methods would "force" a definitive, albeit wrong, result from among the database entries. The analysis did not produce a database organism response; rather, it identified the experimental sample within the group of *S. aureus* strains accurately and according to the unique peptide information consistent with the species level.

5.11   MULTIPLE PROCESSES FOR PROTEIN MIXTURES DISCUSSED

When confronted with multiple processing and concentration steps, confining a complex protein mixture in one container provides quantitative and qualitative benefits. The relative protein signals increase in intensity compared to those in the standard in-solution digestion technique partly because the transfers of the protein mixture to different vials produce sample loss. Furthermore, the filter nature of the vial allowed the low-molecular-weight reactants and by-products to be separated from the processed proteins so as not to interfere with subsequent processing steps. The resulting peptide mixture was easily separated from the trypsin cleavage enzyme. The quantity of peptides and their percent of protein coverage both approximately are doubled, with respect to abundance, from the in-solution digestion method compared to the solution-digestion filter technique introduced herein.

Pure suspensions and mixtures of organisms were amenable to the filter processing method for the capture of the protein to relatively dilute bacterial suspensions and were amenable to the protein and peptide isolation techniques with respect to the identification of the sample. The qualitative and quantitative characteristics of efficient protein extraction were shown by the solution-digestion filter processing method for facilitation in the identification of bacterial samples including mixtures.

This approach also performed well with complex biological mixtures and examples are given in Chapter 7.

6 Computer Software Used for Chemometric and Bioinformatics to Discriminate Microbes

Samir V. Deshpande

CONTENTS

6.1   INTRODUCTION

During MS analysis of the microbial world, huge amounts of data are acquired; therefore, specific algorithms are required to analyze, summarize, and interpret them. In this chapter, commonly used computational methods and algorithms, which are suitable for mining mass spectral data and aim for inferring the presence of the taxonomic position and identity of microbial agents, are reviewed. These algorithms are usually implemented in the form of diverse software tools, which are available commercially or were developed by research groups involved in the MS-based detection and identification of infectious agents. Some of the tools such as ABOid™ can provide detection, identification, and classification results for all three classes of microbes (fungi, bacteria, and viruses) from single MS files.

6.2   BACTERIA CLASSIFICATION USING PATTERN RECOGNITION APPROACHES

The most frequently used unsupervised pattern recognition methods include hierarchical cluster analysis (HCA) and PCA while supervised algorithms such as ANN and PLS-DA represent more recent approaches. During PLS-DA, the principal components (PCs) are rotated to generate latent variables (LVs), which maximize the discriminant power between different classes, in comparison to the total mass spectral variance used in the case of PCA. Therefore, PLS-DA usually gives greatly improved class separation.

6.2.1   Multivariate Linear Least-Squares Regression

A relatively simple approach for processing of MALDI-TOF-MS spectra is represented by a Threat Identification and Detection System (TIDS) software, created by the JHU-APL to automatically detect and characterize mass spectral signature lines (Hayek et al. 1999). Algorithms of the TIDS software handle known, and partially unknown, spectral signatures. For known signatures, an intensity vector was formulated to estimate the similarity of the measured spectrum with a combination of stored library spectra of threat agents. It was achieved by using a multivariate linear least-squares regression of the unknown spectrum to a spectra library. For partially unknown signatures, a Bayesian probabilistic approach has been taken to relate the potentially variable signature of a bacterial threat to likelihoods of chemical composition of bacterial lipids capable of classifying agents on the basis of their chemical (i.e., phospholipid) content (Hayek et al. 1999).

    The mass spectrum can be represented as a two-column table composed of the m/z of detected ions and their intensity. Therefore, a set of similar spectra, that is, spectra composed of the same masses, can be represented by the intensity columns alone and such unique intensity vectors of the biothreat agents were stored in a library.

As an example, Figure 6.1a shows an "unknown" mass spectrum (*B. globigii*) that is to be classified by the processor. Previously stored in the processor's library were signature lines from seven agents derived from a training set of spectra. Three lines in the unknown spectrum correspond to the previously identified *B. globigii* signature and are clearly visible in the range around *m/z* 1100. Figure 6.1b shows the detected

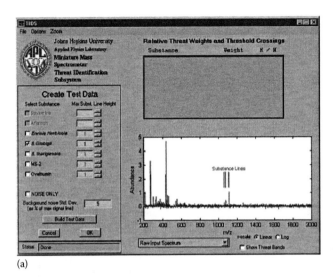

(a)

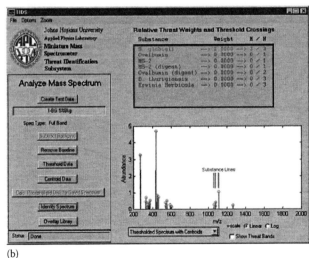

(b)

FIGURE 6.1   Operation of TIDS software. (a) Mass spectrum of an "unknown" (actually the simulant *B. globigii*) presented to the operator by the APL-TOF-MS-TIDS after data collection and prior to signature classification processing. The "Substance Lines" shown here are for the reader's benefit and would not be available to the operator. (b) Correct classification of the unknown in Figure 6.1a by multivariate regression to a library of stored mass spectra signatures derived from a separate training data set. (From Hayek, C.S. et al., *Johns Hopkins APL Tech. Digest*, 20(3), 363, 1999.)

lines overlaid on the threshold spectrum. The probability of false alarms (p_{fa}) for the threshold process was set to 1 in 10,000. Also shown in Figure 6.1b, within the box labeled "Relative Threat Weights and Threshold Crossings" are the results of processing the detected peaks with the identification algorithms. As Figure 6.1a and b illustrates, all three substance lines were detected and correctly classified as belonging to *B. globigii*. Because abundances of the *Bg* "biomarker" ions depend on the growth conditions and differences in sample pretreatment techniques, such as washing a sample with water to remove interfering contaminants prior to analysis, these parameters should be kept constant (Antoine et al. 2004).

In the noiseless case, the problem is to determine optimum weighting coefficients, such that the weighted sum of the intensity vectors in the library best matches the unknown spectrum. Tests with the available data (the four simulants *B. globigii*, *Erwinia herbicola*, MS-2, and ovalbumin in a noiseless, interferenceless environment) showed that each simulant could be readily identified.

6.2.2   HCA

In the HCA method, distances between sample data points are calculated and used to form a 2D plot that represents connectivity and clusters in the data set. This is accomplished by grouping together the samples that are closest in multidimensional space, until all groups are finally merged into one treelike structure known as a dendrogram.

HCA was used to generate bacteria grouping dendrograms based on diverse molecular signatures. For example, Goodacre et al. (1999) used HCA to reveal bacteria groupings based on Py-MS data, while Wilkes et al. (2005a) demonstrated that pattern recognition systems applied to Py-MS data of bacterial cells are able to distinguish strains that differed in serotype, antibiotic resistance phenotype, and PFGE patterns. Many other researchers applied this simple technique to discover clustering of microorganisms based on signatures acquired during MALDI-TOF analyses of whole bacteria cells or through comparisons of genomic/proteomic sequences (Dworzanski et al. 2006; Ecker et al. 2006).

Teramoto et al. (2007) reported the successful classification of *P. putida* strains by MALDI-MS of ribosomal proteins, followed by phylogenetic classification based on cluster analysis of a binary biomatching table constructed from mass spectra. Dworzanski et al. (2006) used peptide sequences identified during proteomic experiments to generate a binary matrix of sequence-to-organism (STO) assignments (SOA). The resulting matrices were further processed to classify and potentially identify an analyzed bacterium using HCA and PCA methods aimed to determine the taxonomic position of an unknown sample. The binary SOA matrix can be represented as a bitmap, i.e., every sequence is treated as a character that can assume only two states (1—present or 0—absent); hence every bacterium is represented as a point in the *n*-dimensional space of peptides, where *n* equals the number of sequence unique peptides. Under these circumstances, squared Euclidean (or city block) distances between bacteria are equivalent to the number of sequences that differentiate them. Such distances were used to calculate a similarity matrix that was analyzed by HCA to reveal groupings among bacterial strains (Dworzanski et al. 2006).

The logic of the data processing workflow used by U.S. Army Edgewood Chemical Biological Center (ECBC) researchers is shown in Figure 6.2 and can be described as follows. During the analysis of an unknown bacterium u, database searches with uninterpreted MS/MS spectra of peptide ions give peptide sequences, which can then be validated using probability criteria. A set of m accepted peptide sequences s_i where $i = 1, 2, 3, ..., m$ can be considered as elements of a column vector $\mathbf{b}_u$ that represents the peptide profile of u composed of m assignments a_{iu} ($\mathbf{b}_u = a_{1u}$, $a_{2u}, a_{3u}, ..., a_{mu}$). Accordingly, STB assignments a_{i1} are elements of a column vector $\mathbf{b}_1$ that represents a peptide profile of a database bacterium assigned as number 1, and in general, assignments a_{ij} are elements of column vectors $\mathbf{b}_j$, where j represents the theoretical proteome of a jth bacterium in the database ($j = 1, 2, 3, ..., n$). All these column vectors form a binary matrix of assignments $A_{m \times (n+1)}$ that can be represented as a virtual array of m peptide sequences assigned to n theoretical proteomes of database bacteria and an unknown microorganism. Each column vector represents a peptide profile of a bacterium, while each row vector represents a phylogenetic profile s_i ($s_i = a_{iu}, a_{i1}, a_{i2}, a_{i3}, ..., a_{in}$) of a peptide sequence. Thus, for each LC-MS/MS analysis, a matrix of STB assignments is created with entries representing the presence or absence of a given sequence in each bacterial theoretical proteome. The absence of a given peptide sequence s_i in a proteome of a microorganism b_j (coded by the jth genome) is represented by zero ($a_{ij} = 0$), while its presence is reflected by the entry of unity ($a_{ij} = 1$).

HCA can be performed using diverse linkage methods (single, complete, Ward's, etc.) and similarity measures (e.g., squared Euclidean distances).

6.2.3 PRINCIPAL COMPONENT ANALYSIS (PCA)

The rationale of the PCA method is the linear transformation of the original variables into a new set of variables called principal components (PCs). They are uncorrelated with each other and may be represented as an orthogonal system of axes, denoted as PC1, PC2, ..., PCn, that, respectively, correspond to a decreasing order of the amount of variance (information) in the data set. For instance, a spatial representation of inter-strain similarities or distances of analyzed bacteria in the data space of PC–PC3 can be presented as a 3D plot that provides the evidence of distinct clusters of points representing bacteria for visual inspection.

PCA of a covariance matrix obtained from an SBA matrix and projections of microorganisms into the data space of the three PCs with the highest eigenvalues was used to evaluate the observed groupings. This approach provided the final groupings of unknown bacteria to database microorganisms (Dworzanski et al. 2006).

6.2.4 ARTIFICIAL NEURAL NETWORKS (ANN)

An artificial neural network (ANN) is a structure composed of interconnected units called artificial neurons with external input and output determined by input–output characteristics and interconnection to other neurons. The most frequently used ANN algorithm for classification of bacterial species based on MS-detected signatures is a multilayer perceptron ANN with a back propagation. The advantage of this approach

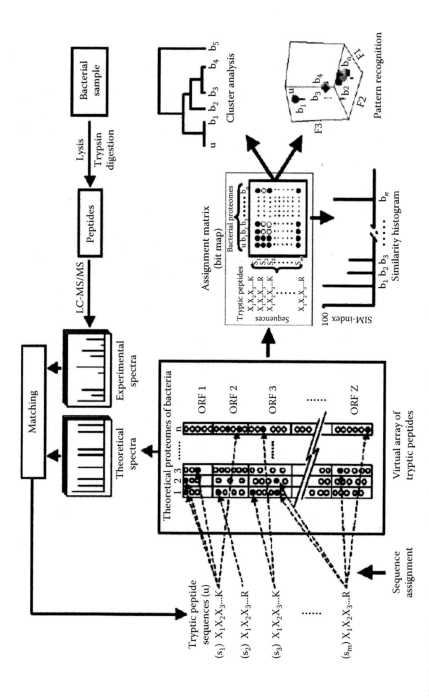

FIGURE 6.2 Schematic representation of sample analysis and data processing workflow for proteomics-based bacteria classification. (From Dworzanski, J.P. et al., *J. Proteome Res.*, 5(1), 76, 2006.)

is that it does not rely on predetermined relationships between variables and identifies patterns in the data through the learning (training) process that resembles synaptic activity of brain neurons by using interneuron connection weights. For example, to identify *Neisseria meningitidis* from other species of this genus and closely related taxa, Lancashire et al. (2005) scaled the data obtained during SELDI-MS analysis between 0 and 1 and used individual *m/z* values with their corresponding relative abundance values as inputs in the input layer, while the output layer consisted of a single node encoded with Boolean representation. During training, the ANN utilized training data set to develop predictive performance and ANN model was validated using a blind validation set. A subset of ions was identified that allowed for the consistent identification of species, correctly classifying more than 97% of samples (Lancashire et al. 2005). The application of ANNs to the analysis of Py-MS spectra has demonstrated that this technique provides a rapid and accurate means of identifying diverse bacterial species (Timmins and Goodacre 2006). For typing bacterial strains by MALDI-MS of whole cells, Bright et al. (2002) developed pattern recognition software that uses an ANN approach to build and search a database.

6.2.5   PARTIAL LEAST-SQUARES DISCRIMINANT ANALYSIS

Soft modeling by partial least-squares discriminant analysis (PLS-DA is a more recent *supervised* pattern recognition approach to the analysis of bacterial MS spectra that attempts to overcome some of the drawbacks observed in PCA. During PLS-DA, the PCs are rotated to generate LVs, which maximize the discriminant power between different classes, not the total mass spectral variance as in PCA, and as such, class separation is greatly improved. Because PLS-DA is a relatively modern soft modeling approach, it has not been extensively evaluated with MS data. Nevertheless, the popularity of this technique for strain discrimination and identification purposes is growing. For example, in the study performed by a Swedish Defence Research Agency, *F. tularensis* subspecies were discriminated using PLS-DA analysis of SELDI-MS data (Lundquist et al. 2005). Pierce et al. (2007) demonstrated the identification of *C. burnetii* cultures using PLS-DA of MALDI-TOF mass spectral peaks acquired for whole cells and reported that this approach allowed for successful differentiation of seven investigated strains.

6.3   TOOLS FOR RAPID ANALYSIS OF MASS SPECTRA ACQUIRED DURING BOTTOM-UP PROTEOMICS EXPERIMENTS

Recent advances in protein identification methodologies that are mainly based on assigning peptide sequences to mass spectra rely on the development of computational approaches that use sequences of known proteins gathered in publicly available databases. In these approaches, spectra of peptides obtained by proteolysis of microbial proteins are compared to theoretical spectra calculated from protein or DNA sequence databases. To correctly categorize the peptides leading to identification of proteins and their sources, that is, bacteria, viruses, or toxins, a huge database has to be processed. Therefore, robust computational capabilities are needed to reduce processing time. There exist a large number of suitable algorithms for matching spectra

(Sadygov et al. 2004), but we will not discuss them in detail. In general, search engines achieve this by matching the specific pattern of peptide fragment ions in the experimental spectra obtained from unknown peptides with theoretically predicated spectra obtained from the protein sequences database. Therefore, the general principle is that these database search engines assign a score between a peptide and spectrum. In this way, virtually each spectrum gives a match to an amino acid sequence in the database. The central issue is how to reliably and automatically control the quality of these assignments to eliminate the false positives. Two aspects have to be dealt with: (1) a pre-filter has to be applied to remove potentially unidentifiable mass spectra, thereby reducing the computational overload and (2) filtering of potentially false positive matches between experimental and theoretical mass spectra. Under these circumstances, a statistical framework is needed, which would estimate the efficiency and significance of the performed filtering expressed in terms of the error rate and sensitivity.

6.3.1   Automated Analysis of MS/MS Spectrum Quality

Modern MSs produce thousands of mass spectra per hour; therefore, the automatic assessment of spectral quality is of paramount importance for efficient filtering of bad quality spectra before the final in-depth processing. In the first step, ion intensities in each spectrum are usually normalized using rank-based intensity normalization rather than relative intensities. In the second step, each spectrum is mapped to the defined feature vectors, which comprise several spectral features like (1) the total number of peaks, (2) the total intensity, (3) the likelihood that two peaks differ by the mass of an amino acid, (4) the normalized intensity of pairs of peaks indicating water loss, etc. To determine boundaries between good and bad quality spectra, Bern et al. (2004) used quadratic discriminant analysis. In this approach, feature vectors of each class were modeled by multivariate Gaussian distributions to find out decision boundaries.

In the other approach, binary classification using support vector machine (SVM) classification has been used. In this approach, SVM incorporating C4.5 and random forest algorithms were applied for analysis of spectra. Nine feature vectors were used: (F1) average intensity of peaks in the spectrum; (F2) standard deviation of the peaks in the spectrum; (F3) total intensity of high peaks on a spectrum; (F4) presence or absence of ammonium ions in the spectrum; (F5) total intensity of peaks; (F6) total intensity of a characteristic fragment ion (y1) derived from the precursor peptide ion; (F7) total intensity of a precursor ion; (F8) total intensity of other characteristic fragment ions; and (F9) score based on naive de novo sequencing the 200 most intense peaks. Thus, the spectrum is filtered using two different statistical algorithms to enable a rigid and automated spectral quality assessment, thereby reducing time and increasing the performance of the tools employed for searching databases (Bern et al. 2004).

6.3.2   Tools for Searching Protein Sequence Databases

The interpretation of amino acid sequences from tandem mass spectra can be labor-intensive and slow; therefore, new algorithms and software tools have been developed to perform the interpretation of tandem mass spectra in a fully automated fashion, thus facilitating high-throughput identification of proteins and their biological sources.

6.3.2.1 SEQUEST

The first algorithm/software application developed to identify proteins by matching MS/MS spectra to database sequences is SEQUEST (Eng et al. 1994) that correlates uninterrupted tandem mass spectra of peptides with amino acid sequences from protein or nucleotide databases (after six frame translation). This way, SEQUEST connects each tandem mass spectrum not only to a protein but also to an organism that synthesized that protein. It uses two scoring functions. The first one is used to rapidly determine a few hundred peptide candidates for each spectrum (preliminary score, Sp) while the second uses cross-correlation of the experimental and theoretical spectra (Xcorr). The preliminary score takes into account the sum of matched fragment ion intensities, the number of total and matched fragment ions as well as the factor that rewards continuity of matching for each ion series (that is, b and y ions). The final score is achieved by converting the expected masses of fragment ions predicted for any database peptide into a theoretical spectrum and by computing a cross-correlation between the theoretical spectrum and the experimental spectrum. In addition, SEQUEST exports the normalized difference between the best and the second best scores (ΔCn) that is useful to determine match uniqueness relative to near misses for a given database.

6.3.2.2 Mascot

Mascot has been developed by Matrix Science, UK (http://www.matrix science.com), and was described in 1999 (Perkins et al. 1999); nevertheless, the scoring used by this software application has never been published or patented. It is known that it involves the selection of two fragment ion types, where most fragment ion matches are observed, and a probability-based score computed by using these fragments. However, preprocessing of the experimental mass list is also a part of this algorithm and the final score is the negative logarithm of a p-value. The probability-based scoring algorithm has a number of advantages because a simple rule can be used to judge whether a match result is significant or not. This is particularly useful in guarding against false positives. Furthermore, Mascot scores can be compared with those from other search engines and search parameters can be readily optimized by iteration.

6.3.2.3 X!Tandem

X!Tandem is a fast protein database search engine available from the Global Proteome Machine Organization. X!Tandem is an open source software that automatically searches for missed cleavages, semi-tryptic peptides, PTM, and point mutations. X!Tandem was created by Craig and Beavis (2004) of the Manitoba Proteomics Center as an open source alternative to SEQUEST, Mascot, and similar search engines. X!Tandem, like Mascot and SEQUEST, compares each spectrum to all likely candidate peptides in a protein database. One of X!Tandem's strengths is its automatic search for modified peptides. X!Tandem works by matching acquired MS/MS spectra to a model spectrum based on peptides in a protein database. The model spectrum is very simple, based on the presence or absence of y and b ion. Only matching spectral peaks are considered, while any peaks that do not match, in either the model or acquired spectra, are not used. Therefore, the acquired spectrum is simplified to only those peaks that are similar to the peaks in the model spectrum. The hyperscore used by X!Tandem is much faster to calculate than SEQUEST's Xcorr. This allows X!Tandem to build histograms that allow the calculation of E-values. For nonspecific searches, this software performs 200 to 1000 times faster when assessing the presence of frequently occurring peptide modifications like oxidation, deamidation, and phosphorylation.

6.3.3 BIOINFORMATICS FOR PROTEOMICS

The development of bioinformatic tools for proteomic applications has focused on filtering algorithms and statistical methods to validate proteomics data. Today, there are numerous Internet sites hosting open source projects and the most known are Bioinformatics.org (over 300 projects), sourceforge.net (over 900 projects), and Open Bioinformatics Foundation (hosts toolkits like BioJava, BioPerl, and BioPython) (Stajich and Lapp 2006). One of the latest distributions available is NEBC Bio-Linux developed by the NERC Environmental Bioinformatics Centre (Field et al. 2006). Spjuth et al. (2007) developed an open source software application called Bioclipse that provides a complete and extensile tool kit for chemoinformatics and bioinformatics. This java-based Rich Client Platform (RCP) software integrates multiple frameworks and components like Biojava, Chemistry Development Kit (CDK), JChemPaint, Spectrum, and many more. Bioclipse is an advanced powerful workbench that offers resources for (i) molecules, proteins, sequences, and spectra, (ii) 2D editing and 3D visualization, (iii) networked servers, clusters, and databases, and (iv) local file systems and devices (e.g., printers).

The software package, Multi-Q, was developed by Lin et al. (2006) and provides converters for spectral data files from major manufacturers of MS, that is WIFF (Applied Biosystems), RAW (Thermo Scientific, Waters), and BAF (Bruker Daltonics), and reduces them to an mzXML file format. In addition, Multi-Q accepts search results from different engines (Mascot, SEQUEST, X!Tandem) in CVS or HTML formats. Moreover, Multi-Q allows the user to define their own filtering and statistical module to correct: (i) low-quality MS/MS spectra; (ii) random and systematic errors in detection responses; and (iii) sample handling.

In 2005, Gärdén et al. (2005) introduced a new open source client server called PROTEIOS that provides additional plug-ins for protein identification, data viewing, and tool analysis. Furthermore, this application can import mzData, mzXML, and sample generation/sample-processing parts of Proteome Experimental Data Repository (PEDRo) (Taylor et al. 2003). PROTEIOS is implemented in java, SQL, and is platform-independent. Rauch et al. (2006) have recently introduced an open source Computational Proteomics Analysis System (CPAS). In contrast to other software management systems (such as SBEAMS and PRIME), CPAS includes the following features: (a) multiple standard file formats such as FASTA, mzXML, and pepXML; (b) an experiment annotation to track and organize biological experiments and view the workflow; and (c) tools to promote collaborative projects. The XML is a language that facilitates the sharing of data across different information systems and the implementation of this standard facilitates information exchange and sharing of microbial proteomics data.

6.3.3.1 Trans-Proteomic Pipeline

The Trans-Proteomic Pipeline (TPP, Figure 6.3) is a collection of integrated tools for processing database search results performed with product ion spectra of peptide ions. TPP has been developed at the Institute of System Biology's (ISB) Seattle Proteome Center (SPC) and includes (a) data input module, (b) peptide assignment and validation module, (c) protein assignment and validation, (d) quantification module, and (e) interpretation module.

TPP uses the mzXML format as its input module and is easily extensible to additional search engines and analysis modules because this format is now an industry standard for generating and converting data from MS in a generalized XML format.

The second module incorporates the data searching tools like SEQUEST and Mascot as discussed earlier in this chapter. The results derived from this module are fed into the "validation module," which uses PeptideProphet algorithm (Keller et al. 2002) to validate peptide sequence assignments provided by database search engines through statistical analysis of matches. PeptideProphet is a robust and accurate software and uses the expectation maximization algorithm for validation of database searching results. By employing database search scores and other information, the system learns to distinguish correctly from incorrectly assigned peptides and computes for each peptide assignment a probability of being correct. By using the probabilities computed from the model, one can achieve much higher sensitivity for any given error rate compared to the results of using conventional filtering criteria. The method enables high-throughput analysis of proteomics data by eliminating the need to manually validate database search results. In addition, it can facilitate the benchmarking of various experimental procedures and serve as a common standard by which the results of different experimental groups can be compared.

Since MS/MS spectra are produced by peptides, and not proteins, there is a need for an additional statistical model for validation of the identifications at the protein level. Hence, the results are then passed through ProteinProphet (Nesvizhskii et al. 2003), which is a statistical model for validation of peptide identification at the protein level. It uses a list of peptides assigned to MS/MS spectra and corresponding

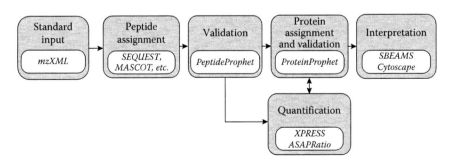

FIGURE 6.3 TPP data flow.

probabilities that those peptide assignments are correct. Different peptide identifications corresponding to the same protein are combined together to estimate the probability that their source protein is present in the sample. This protein grouping information is then employed to adjust the individual peptide probabilities, thus making the approach more discriminative. Finally, interpretation and quantification modules are used to consolidate the results.

6.4 DEVISING A BIOINFORMATIC PLATFORM FOR IDENTIFICATION AND COMPARATIVE ANALYSIS OF MICROBIAL AGENTS USING PEPTIDE MS/MS (ABOid™)

Currently, more than 2000 bacteria have been fully sequenced and hundreds of sequencing projects are in progress. Completely sequenced genomes provide amino acid sequence information of every protein potentially expressed by these organisms. Hence, the combination of this resource with MS technologies capable of identifying amino acid sequences of proteins enables one to design new procedures for the classification and identification of bacteria based on querying proteomic sequences. Although the MS/MS-based sequencing of peptides by using database search engines or by de novo sequencing of peptides are common practices (Aebersold et al. 2003), it is still a challenging task to translate the raw data generated from MS/MS experiments into a biologically meaningful and easy-to-interpret results suitable for identification and classification of microorganisms with high confidence.

Researchers in the U.S. Army ECBC's Point Detection Team developed a suit of bioinformatics tools for the rapid classification and potential identification of bacteria based on the peptide sequence information generated from LC-ESI-MS/MS analysis of tryptic digests of bacterial proteins. An integrated and automated software application that uses such an approach for rapid bacteria identification is referred to as Agents of Biological Origin Identification (ABOid™) which is an evolved refinement of BACid, the progenitor. ABOid™ determines identification results by searching product ion spectra of peptide ions against a custom protein database, created by commercially available software (e.g., SEQUEST), into a taxonomically meaningful and easy-to-interpret output. To achieve this goal, a protein database is constructed in a FASTA format that consists of theoretical proteomes derived from all fully sequenced bacterial genomes. Each protein sequence in this database was supplemented with information on a source organism, chromosomal position of its respective open reading frame (ORF), and linked to the microbial taxonomy database.

ABOid™ analyzes SEQUEST search results files and computes probabilities that peptide sequence assignments to product ion mass spectra (MS/MS) are correct (Figure 6.4). In the next step, these probability scores are used to filter out low confidence peptide assignments by selecting a suitable value for a "threshold cutoff" parameter. Further, ABOid™ calculates assignment error for the accepted set of spectrum-to-sequence matches and uses these high confidence assignments to generate a STB binary matrix of assignments (SBA). These SBA matrices show validated peptide sequences, which are differentially present or absent in various strains being compared and could be visualized as bitmaps (e.g., Figure 6.5a). In addition, the

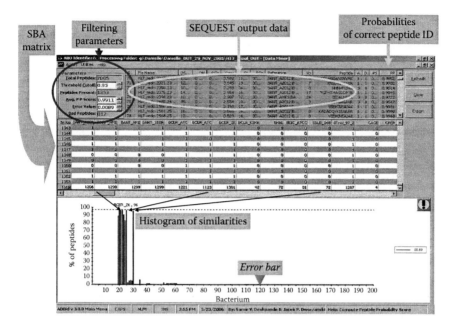

FIGURE 6.4   Screen capture image displaying filtered data and histogram of similarities.

number of peptide assignments to database organisms can be displayed as similarity histograms (Figure 6.4). Finally, the bacterial classification and identification algorithm uses assignments of organisms to taxonomic groups based on an organized scheme that begins at the phylum level (Figure 6.5b) and follows through classes, orders, families, and genera down to strain level.

In Figure 6.5, a data analysis pathway used for analysis of a bacterial mixture composed of *E. coli* and *B. cereus* cells is presented. In this example, the SOA matrix of assignments (Figure 6.5b) was analyzed by computing assignments to merged proteomes that comprise bacteria grouped into "super-proteomes" of 13 phyla represented in the database. The results shown in Figure 6.5b indicate that 98 unique sequences were assigned to the phylum Proteobacteria while 99 were assigned to Firmicutes, thereby confirming the presence of a mixture of bacteria and allowing the classification of these organisms on the lower taxonomic levels. The assignment submatrices was analyzed separately and the results are shown in Figure 6.5c and d as dendrograms representing results of cluster analyses.

An experimentally determined error rate associated with a set of accepted peptide sequences is used in testing the significance of any observed matching discrepancies. During the classification process, the algorithm applies decision criteria based on the significance of computed similarities between a test sample and taxons at each classification level. In the case of pure culture analysis, similarities between an unknown sample and database bacteria for construction of similarity histograms are calculated by using a Jaccard's coefficient. In the first test, the difference between similarity coefficient values for the two highest scoring taxa is compared to the known error rate for the data set ("noise level"). When it exceeds a preset threshold value (e.g., $S/N = 3$),

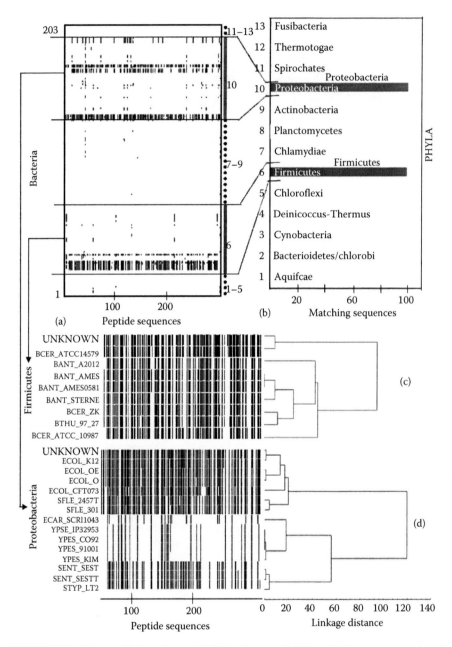

FIGURE 6.5 Data analysis pathway. (a) Virtual array of 289 peptide sequences assigned to proteomes of 209 bacteria; (b) histogram of the number of matching sequences assigned to "super-proteomes" of 13 phyla obtained by merging database bacteria according to their taxonomic position; (c) cluster analysis of a STB assignment matrix for Firmicutes; and (d) cluster analysis of a STB assignment matrix for Proteobacteria. (From Deshpande S.V. et al., *J. Chromat. Separation Techniq.,* 1(1), S5, 2011.)

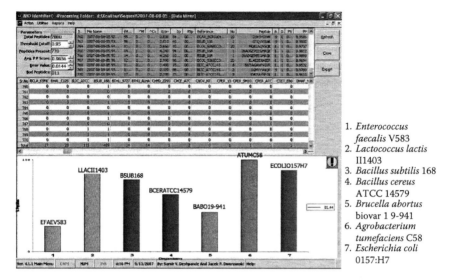

1. *Enterococcus faecalis* V583
2. *Lactococcus lactis* Il1403
3. *Bacillus subtilis* 168
4. *Bacillus cereus* ATCC 14579
5. *Brucella abortus* biovar 1 9-941
6. *Agrobacterium tumefaciens* C58
7. *Escherichia coli* 0157:H7

FIGURE 6.6   Screen capture image displaying results of bacterial mixture analysis.

the difference is deemed to be significant and a taxon with the highest similarity is considered as a potential candidate for the correct sample classification. To verify the correctness of the earlier classification test, a second test was devised, which is based on the following premises. Namely, if the first test is correct, then peptides assigned to other taxa at the same classification level (i.e., phylum, class, order, etc.) should comprise only subsets of those matching the highest scoring taxon. Conversely, if the first test is incorrect, it means that some peptides have unique sequences for other taxa. However, the presence of such unique peptides is considered as statistically significant only if the number of sequences unique for a given taxon exceeds the noise level (S/N = 3). In the case of bacterial mixtures, only the second test is used.

The results of the data mining process applied to the analysis of bacterial mixture composed of seven bacterial strains are displayed in Figure 6.6. In this case, all seven strains analyzed by LC-MS/MS and BACid were correctly identified. The ABOid™ algorithm was developed using MATLAB® and Microsoft Visual Basic, while the Phylogenetic Classification module is incorporated in the Data Analysis and data warehousing software designed and developed in-house. In summary, ABOid™ examines a large number of peptide sequences from a single LC-ESI-MS/MS analysis of a bacterial protein digest for rapid classification and identification of pure bacterial cultures as well as microbial mixtures.

6.5   IDENTIFICATION OF MICROBES USING PCR-MS

A key element of the PCR-MS-based approach for pathogen identification and strain typing is the assembly of a database containing genomic information for thousands of organisms. This information associates base compositions and masses of potential amplicons with primer pairs for each organism and is carefully curated through sequence alignments and extensive primer scoring for target organisms.

The software developed by Ibis scientists (GenX) uses raw MS data to establish the exact mass of each strand of the amplified genomic segment during the PCR procedure. The number of consistent base compositions found for a given mass of the amplicon is always relatively high, even by using high-resolution MS that are capable of providing mass vincertainty as low as 1 ppm. For instance, for a single DNA strand (M_r of 37,374.266 Da) at the 1 ppm mass accuracy level the number of possible base compositions exceeds 100, while at the 10 ppm mass uncertainty this number approaches 1000. However, even at the 20 ppm accuracy level the base composition of the amplicon pair is constrained to a single, unique base composition because the base composition for one strand must be complementary to that for the second. Finally, results are "triangulated" through searching a database containing genomic information to determine organism assignments (Ecker et al. 2006).

6.6   ABOid™: SOFTWARE FOR AUTOMATED IDENTIFICATION AND PHYLOPROTEOMICS CLASSIFICATION OF TANDEM MASS SPECTROMETRIC DATA

6.6.1   INTRODUCTION

A suite of bioinformatics algorithms called ABOid™ (Deshpande et al. 2011) was developed for an automated identification and classification process for microbes based on the comparative analysis of protein sequences. This application uses sequence information of microbial proteins revealed by MS-based proteomics for identification and phyloproteomics classification. The algorithms transform results of searching product ion spectra of peptide ions against a protein database, performed by commercially available software (e.g., SEQUEST), into a taxonomically meaningful and easy-to-interpret output. To achieve this goal, we constructed a custom protein database composed of theoretical proteomes derived from all fully sequenced bacterial genomes (1204 microorganisms as of August 25, 2010) in a FASTA format. Each protein sequence in the database is supplemented with information on a source organism and chromosomal position of each protein-coding ORF is embedded into the protein sequence header. In addition, this information is linked with a taxonomic position of each database bacterium.

ABOid™ analyzes SEQUEST search results files to provide the probabilities that peptide sequence assignments to a product ion mass spectrum (MS/MS) are correct and uses the accepted spectrum-to-sequence matches to generate a STO matrix of assignments. Because peptide sequences are differentially present or absent in various strains being compared, this allows for the classification of bacterial species in a high-throughput manner. For this purpose, STO matrices of assignments, viewed as assignment bitmaps, are next analyzed by a ABOid™ module that uses phylogenetic relationships between bacterial species as a part of decision tree process, and by applying multivariate statistical techniques (PC and cluster analysis), to reveal relationship of the analyzed unknown sample to the database microorganisms. This bacterial classification and identification algorithm uses assignments of an analyzed organism to taxonomic groups based on an organized scheme that begins at the phylum level and follows through classes, orders, families, and genus down to the strain level.

We have developed a suite of bioinformatics algorithms for automated identification and classification of microbes based on comparative analysis of protein sequences. This application uses sequence information of microbial proteins revealed by MS-based proteomics for identification and phyloproteomics classification. The algorithms transform results of searching product ion spectra of peptide ions against a protein database, performed by commercially available software (e.g., SEQUEST), into a taxonomically meaningful and easy-to-interpret output. To achieve this goal, we constructed a custom protein database composed of theoretical proteomes derived from all fully sequenced bacterial genomes (1204 microorganisms as of August 25, 2010) in a FASTA format. Each protein sequence in the database is supplemented with information on a source organism and chromosomal position of each protein-coding ORF is embedded into the protein sequence header. In addition, this information is linked with the taxonomic position of each database bacterium.

6.6.2 SAMPLE PROCESSING FOR ABOID™ EVALUATION

Bacterial cells (*E. coli* K-12 and *B. cereus* ATCC 14579) were grown and processed before LC-MS analysis as described previously (Dworzanski et al. 2004b). Bacteria were lysed by sonication and proteins extracted from ruptured cells were denatured with urea, reduced with DTT, alkylated with iodoacetamide, and digested by trypsin.

A mixture of peptides obtained by trypsin digestion of their cellular proteins were separated on a C18 column (100 μm i.d. × 100 mm) and analyses of electrosprayed peptide ions were carried out using an ion trap MS (Thermo Scientific, San Jose, CA). Each MS data acquisition cycle consisted of a full-scan MS over the mass range m/z 400–1400, followed by data-dependent MS/MS scans over m/z 200–2000 on the five most intense precursor ions from the survey scan.

6.6.3 DATA PROCESSING

MS/MS spectra were searched with TurboSEQUEST (Bioworks 3.1; Thermo Scientific, San Jose, CA) against a database constructed from FASTA-formatted theoretical proteomes downloaded from a National Center for Biotechnology (NCBI) server, which are predicted from fully sequenced bacterial genomes. SEQUEST output files were processed with ABOid™.

6.6.3.1 Protein Database and Database Search Engine

A protein database was constructed in a FASTA format using the annotated bacterial proteome sequences derived from fully sequenced chromosomes of 1125 bacteria, including their sequenced plasmids (as of August 25, 2010). A PERL program was written to automatically download these sequences from the National Institutes of Health NCBI site (http://www.ncbi.nlm.nih.gov). Each database protein sequence was supplemented with information about a source organism and a genomic position of the respective ORF embedded into a header line. The database of bacterial proteomes was constructed by translating 8,323,020 putative protein-coding genes and consists of 2,521,160,789 amino acid sequences of potential tryptic peptides obtained by the in silico digestion of all proteins (assuming up to two missed cleavages). The experimental MS/MS spectral data of bacterial peptides were searched by a SEQUEST algorithm against this protein database (Eng et al. 1994).

6.6.4 PROGRAM OPERATION

ABOid™ is a suite of algorithms developed in-house using Microsoft Visual Basic (VB).NET and PERL to analyze bacterial similarities and their identification from a virtual array of STO matrix of assignments. A data flowchart is shown in Figure 6.7.

6.6.5 DBCURATOR

The first module (dbCurator) written in Perl downloads the microorganism sequences and edits the header information of each protein. This new theoretical proteome of a microorganism is appended in the existing flat file that is saved as FASTA format; dbCurator also updates the microorganism relational database (My ABOid) created using MySQL (http://www.mysql.com/) with the microorganism information like name, strain, sequencing center, and other available data related to each bacterium. ABOid™ utilizes two databases: flat file FASTA format database used as the reference and My ABOid, which is a central repository database.

6.6.6 BACDIGGER

The second module (BacDigger) is designed to analyze the SEQUEST output files. The function of this module is to retrieve sequence matches to the in-house reference bacterial proteomes based on the identity of the peptides determined by SEQUEST and to obtain values for the matching parameters (Eng et al. 1994) like Xcorr, ΔCn, Sp, RSp, ΔM, and number of amino acids in the peptide sequence identified assigning each peptide sequence, a probability score determined by running PeptideProphet (Keller et al. 2002) algorithm to validate the SEQUEST peptide sequence assignments. Using the information contained in the reference of each output file, a STO binary matrix of assignments is created. This matrix of assignments, generated using raw results, is archived in a comma separated file format (CSV) for audit. Based on probability values, determined by a PeptideProphet algorithm that a sequence was correctly identified, a user-specified threshold is applied to elements of the STO matrix of assignments to filter out low probability matches. This new "extracted" STO matrix of assignments is also saved in a CSV format. In addition, BacDigger removes duplicate modules to display the most probable taxonomic position of studied microorganisms and provides a user-friendly display of results.

The application's graphical user interface (GUI) is developed in Microsoft Visual Basic.NET (http://msdn.microsoft.com/vbasic); data processing algorithms are developed using Microsoft C++ and PERL (http://www.activestate.com/Products/ActivePerl). Statistical analysis is performed using R 2.4.1 and Statistica (Statsoft, Inc., Tulsa, OK). MySQL Server is used to archive the data and results.

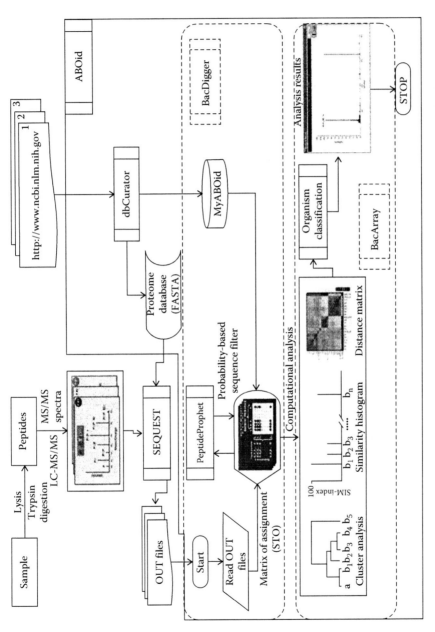

FIGURE 6.7   Flow chart of the ABOid™ application (From Deshpande, S.V. et al., *J. Chromat. Separation Techniq.*, S5, 2011.)

6.6.7 RESULTS

MS/MS spectra of peptide ions generated during the ESI process of tryptic peptides derived from bacterial proteins were searched against the protein database with SEQUEST and the output files were processed by ABOid™ (Figure 6.8). Amino acid sequences of peptides were validated using probability scores generated by PeptideProphet. A set of 289 accepted peptide sequences ($p>0.98$) from the data set results. This set of 289 peptides includes a subset of peptides that are unique and these unique peptides are used as elements to form a row vector b_u in the BacArray.

The third module (BacArray) takes the STO matrices of assignments and displays numerical values in the form of color bitmaps ("virtual arrays") as shown in Figure 6.9. During this process, BacArray rearranges assignments by grouping database bacteria according to their taxonomic positions by using the relevant information stored in My ABOid. This allows for interactive browsing of sequence assignments, which can be further validated as they are dynamically linked with NCBI protein databases for blasting the sequences of interest.

The BacArray communicates with external statistical libraries to apply multivariate statistical techniques like PC and cluster analysis to the STO matrices of assignments. In addition, it can generate combined reports of such analyses, thus enabling the peptide profile of unknown (u). Accordingly, STO assignments a_{1i} are elements of a row vector b1 that represents a peptide profile of a database bacterium assigned as number 1, and in general assignments a_{ij} are elements of row vectors b_i, where i represents the theoretical proteome of a ith bacterium in the database ($i=1, 2, 3, ...,$ 203). All these row vectors form a matrix of assignment $A(m+1) \times n$ that is visualized in Figure 6.8a as a virtual array of $n=289$ peptide sequences assigned to $m=202$ theoretical proteomes of database bacteria and an unknown microorganism (or their mixture). Conversely, each column vector represents a phylogenetic profile s_j of a peptide sequence. Thus for each MS/MS analysis, a binary matrix of assignments A is created with entries representing the presence or absence of a given sequence in each theoretical proteome of database microorganism. Similar b profiles indicate a correlated pattern of relatedness that in the majority of cases reflects the presence of identical sequences among orthologs or other functional gene segments. The method predicts that peptide sequences b_u derived from cellular proteins of an unknown bacterium u are most likely to be similar or even identical with a reference database bacterial strain b_i represented by a vector b_i with a highest number of nonzero elements.

The STO matrix of assignments was next analyzed by computing sequence assignments to merged proteomes that comprise bacteria grouped into "super-proteomes" of 13 phyla represented in the database. The results shown in Figure 6.8b indicate that 98 unique sequences were assigned to the phylum Proteobacteria while 99 were assigned to Firmicutes, thus, confirming the presence of a mixture of bacteria and allowing the classification of these organisms on the phylum level. The STO assignment submatrices were further analyzed separately and the results obtained are shown in Figure 6.8c. Figure 6.8d shows dendrograms representing results of cluster analyses that are accompanied by bar graphs that visualize peptide profiles of the test samples ("unknown") and the most similar database strains revealed by cluster analysis.

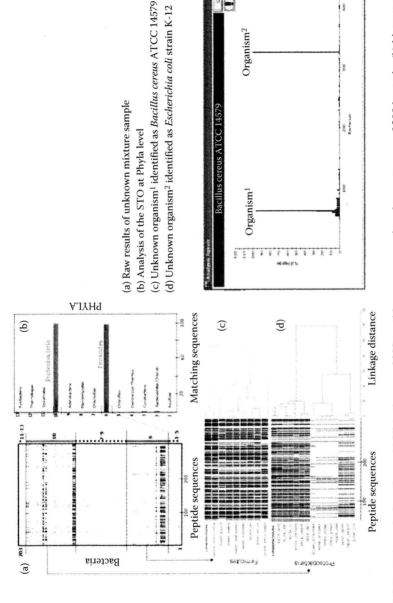

(a) Raw results of unknown mixture sample

(b) Analysis of the STO at Phyla level

(c) Unknown organism[1] identified as *Bacillus cereus* ATCC 14579

(d) Unknown organism[2] identified as *Escherichia coli* strain K-12

FIGURE 6.8   Data analysis pathway. (a) Virtual array of 289 peptide sequences assigned to proteomes of 209 bacteria; (b) histogram of the number of matching sequences assigned to "super-proteomes" of 13 phyla obtained by merging database bacteria according to their taxonomic position; (c) cluster analysis of a STO assignment matrix for Firmicutes; and (d) cluster analysis of a STO assignment matrix for Proteobacteria. (From Deshpande, S.V. et al., *J. Chromat. Separation Techniq.*, S5, 2011.)

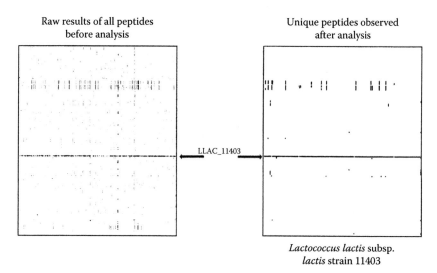

Raw results of all peptides before analysis

Unique peptides observed after analysis

LLAC_11403

Lactococcus lactis subsp. *lactis* strain 11403

FIGURE 6.9   Virtual array bitmap. (From Deshpande, S.V. et al., *J. Chromat. Separation Techniq.*, S5, 2011.)

Cluster analysis provides a way to determine groups based on similarity within a data set, performs PCA to derive parsimony, and reduces the dimensionality to measure the variation among the proteins identified from different strains of a same species. Hierarchical clustering was performed using furthest neighbor (complete) linkage with squared Euclidean distances as the similarity metric and was used as an exploratory tool to examine relationships of a test microorganism with the database bacteria. Cluster analyses were performed automatically by linking STA Cluster library from Statistica to the BacArray module.

In Figure 6.8c, one can observe two main clusters: the first cluster groups the "unknown" with a database strain *B. cereus* ATCC 14579, while the second cluster comprises diverse *Bacillus* strains classified as *B. cereus*, *B. anthracis*, and *B. thuringiensis*.

In Figure 6.8, the test sample forms a subcluster with a database *E. coli* K-12 strain because they differ only by two peptide sequences. This subcluster is grouped with other *E. coli* and *S. flexneri* strains into a cluster that is substantially different in comparison to the next closest cluster that comprises *Salmonella* and *Yersinia* strains.

Figure 6.9 shows a virtual array plot of a single organism identified as *Lactococcus lactis* subsp. *lactis* strain I1403 from the raw data before processing and unique peptides observed after data analysis. Figure 6.10 shows the algorithm-generated results for seven organism mixture identified as (1) *B. cereus* ATCC 14579, (2) *S. aureus*, (3) *S. pyogenes*, (4) *Burkholderia thailandensis*, (5) *E. coli* strain K-12, (6) *S. enterica*, and (7) *P. aeruginosa* strain PA01. All the samples used in the analysis had been double-blind bacterial samples. Further utility of ABOid™ on double-blind bacterial samples can be found in a *Chromatography Separation Techniques* publication (Deshpande et al. 2011). Figure 6.11 is a screenshot of the application.

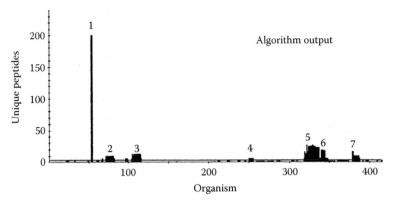

FIGURE 6.10   Results of blinded study showing results of the seven organism mixture. (From Deshpande, S.V. et al., *J. Chromat. Separation Techniq.*, S5, 2011.)

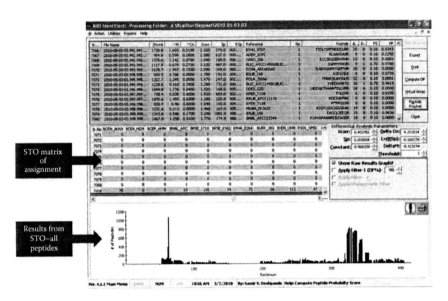

FIGURE 6.11   Screenshot of the ABOid™ application. (From Deshpande, S.V. et al., *J. Chromat. Separation Techniq.*, S5, 2011.)

6.6.8   CONCLUSION

The results of applying ABOid™ for analysis of a bacterial sample composed of a mixture of *E. coli* K-12 and *B. cereus* ATCC 14579 strains demonstrate that MS-based proteomic approach, combined with ABOid™ for analysis of SEQUEST output files, allows for automated assignment of analyzed organisms to taxonomic groups. Moreover, ABOid™ reveals genome-traced relatedness between bacteria that is suitable for fast and reliable classification and even identification of bacteria up to the strain level. Therefore, the application of this algorithm for analyses of proteomics data constitutes a new method that may function as a strong complement to DNA-based approaches to comparing bacterial genomes.

ABOid™ is capable of revealing the identification of microbial mixture contents. The software application does not require prior knowledge of the sample and can be applied to pure cultures and mixtures. It allows for strain-level identification based on comparative analysis of protein sequences and the unsequenced bacterial strains not in our database are identified to their close-neighbor species. Statistical and visualization tools of the software allow its utilization by nonspecialist end users.

7 Applications

Charles H. Wick

CONTENTS

7.1 INTRODUCTION

Little is the need to demonstrate the capability of a one-step solution to the detection and identification of microbes by MSP using ABOid™ software. The following examples are demonstration enough. Some of these have been published in the open peer-reviewed literature in shorter versions to confirm the acceptance of this approach by the scientific community. The honeybee paper has had more than 70,000 views and when published was included on the front page of more than 100 newspapers worldwide. The obvious conclusion after reading these examples is that this methodology/approach is indicated where accurate microbe detection is desired.

7.2 DOUBLE-BLIND CHARACTERIZATION OF NON-GENOME-SEQUENCED BACTERIA BY MSP

One reason to develop a one-step method to detect microbes is the possibility of epidemic. This creates a need for technologies that can detect and accurately identify pathogens in a near-real-time approach. One technology that meets this capability is a high-throughput MS-based proteomic approach. This approach utilizes the knowledge of amino acid sequences of peptides derived from the proteolysis of proteins as a basis for reliable bacterial identification. To evaluate this approach, the tryptic digest peptides generated from double-blind biological samples containing either a single bacterium or a mixture of bacteria were analyzed using LC-MS/MS. Bioinformatic tools that provide bacterial classification were used to evaluate the proteomic approach. Results show that bacteria in all of the double-blind samples were accurately identified with no false-positive assignment. The MSP approach showed strain-level discrimination for the various bacteria employed. The approach also characterized double-blind bacterial samples to the respective genus, species, and strain levels when the experimental organism was not in the database due to its genome not having been sequenced. One experimental sample did not have its genome sequenced, and the peptide experimental record was added to the virtual bacterial proteome database. A replicate analysis identified the sample to the peptide experimental record stored in the database. The MSP approach proved capable of identifying and classifying organisms within a microbial mixture.

The detection and accurate identification of pathogens of biological origin are of great importance to the armed forces and civilian sectors. Achieving these tasks is vital in the response to manmade or natural biothreat attacks in a proper and efficient manner to minimize the outbreak of epidemic cases. Several approaches reported in the literature have addressed the detection and identification of microorganisms based on the characterization of metabolites (Akoto et al. 2005; Moe et al. 2005) and genomic contents of bacterial cells (Mayr et al. 2005). In these studies, the genomic sequence similarities generated from PCR were used to group bacteria at the genus/species level (Woese et al. 1985). Prior knowledge of the sample, or the targeting of one or a group of biological substances, is required in PCR techniques for proper primer utilization. However, proteins constitute greater than 60% of the dry weight of microorganism cellular components (Cain et al. 1994; Holland et al. 1996; Duché et al. 2002; Hesketh et al. 2002; Tan et al. 2002) and could provide in-depth information for the bacterial differentiation of species and their strains. Moreover, advancements in MS ionization, detection methods, and data processing make MS a suitable analytical technique for the differentiation of microorganisms (Chen et al. 2001; Demirev et al. 2001; Craig and Beavis 2004).

Using MS techniques for bacterial differentiation relies on the comparison of the proteomic information generated from the analysis of either intact protein profiles (top down) or the product ion mass spectra of digested peptide sequences (bottom up) (Washburn et al. 2001; Warscheid and Fenselau 2003). For top-down analysis, bacterial differentiation is accomplished through the comparison of the MS data of intact proteins to those of an experimental mass spectral database containing the mass spectral fingerprints of the studied microorganisms (Demirev et al. 2001; Craig and Beavis 2004). Conversely, bacterial differentiation using the product ion mass spectral data of digested peptide sequences is accomplished through the utilization of search engines for publicly available sequence databases to infer identification (Xiang et al. 2000; Warscheid and Fenselau 2004). Several peptide-searching algorithms (i.e., SEQUEST and Mascot) have been developed to address peptide identification using proteomics databases that were generated from either fully or partially genome-sequenced organisms (Eng et al. 1994; Perkins et al. 1999; Craig and Beavis 2004). Thus, our approach is based on a cross-correlation between the generated product ion mass spectra of tryptic peptides and their corresponding bacterial proteins resident in an in-house comprehensive proteome database from online databases of the sequences of microorganism genomes (Yates and Eng 2000). Recent developments in the microbial differentiation field have focused on improving the selectivity of MS data processing. The product ion mass spectrum-SEQUEST approach was reported for the identification of specific bacteria using a custom-made, limited database of sequences (Keller et al. 2002; VerBerkmoes et al. 2005). Another approach used ORF translator programs to predict possible protein sequences from all probable ORFs and correlate them with the genomic sequences to establish an identification of microorganisms (Chen et al. 2001). This approach did not show advantages over the product ion mass spectrum method with regard to strain-level discrimination (Wolters et al. 2001). However, a recent advancement in proteomic approaches to bacterial differentiation reported a hybrid approach combining protein profiling and sequence database searching using accurate mass tags (Lipton et al. 2002; Norbeck et al. 2006). This approach was used to probe defined mixtures of bacteria to evaluate its capabilities.

Alternatively, our approach is based on a cross-correlation between the product ion spectra of the tryptic peptides and their corresponding bacterial proteins derived from an in-house comprehensive proteome database from genome-sequenced microorganisms (Dworzanski et al. 2004, 2006; Deshpande et al. 2011). The exploitation of this proteome database approach allowed for a faster search of the product ion spectra than that using genomic database searching. Also, it eliminates inconsistencies observed in publicly available protein databases due to the utilization of nonstandardized gene-finding programs during the process of constructing the proteome database. The proposed approach uses an ensemble of bioinformatic tools for the classification and potential identification of bacteria based on the peptide sequence information. This information is generated from the LC-MS/MS analysis of tryptic digests of bacterial protein extracts and the subsequent profiling of the sequenced peptides to create a matrix of STB assignments. This proteomic approach is an unsupervised approach to reveal the relatedness between the analyzed samples and the database of microorganisms using a binary matrix approach. The binary matrix is analyzed using diverse visualization and multivariate statistical techniques for bacterial classification and identification.

This study investigated the capability of the aforementioned MS-based proteomic approach to identify biological agents using double-blind (hereafter referred to as blind) samples that consisted of various microorganisms of interest to civilian and military installations. The present study included category A biological agents, mixtures of organisms, and negative controls without prior knowledge of the identity of the microorganisms. The in-house database consists of 881 microbial genomes as of May 2, 2009. The identification process for all samples revealed that several samples consisted of a mixture of bacterial species. The results of the blind studies showed a promising outlook for applying this MS-based proteomic approach to the classification of unknown bacterial mixtures at the species and strain level depending on the availability of complete genome sequences.

7.2.1 MATERIALS AND METHODS

7.2.1.1 Materials and Reagents

ABC, DTT, urea, ACN (HPLC grade), and formic acid were purchased from Burdick and Jackson (St. Louis, MO). Sequencing grade–modified trypsin was purchased from Promega (Madison, WI).

7.2.1.2 Biological Sample Preparation

Twenty-one blind biological samples were prepared by streaking cells from cryopreserved stocks onto appropriate agar. B. subtilis, B. thuringiensis, S. aureus, E. faecalis, and P. aeruginosa were streaked onto tryptic soy agar (TSA; catalog number CM100; Culture Media and Supplies, Oswego, IL) plus 5% sheep blood. B. thailandensis and Clostridium phytofermentans ISDg were streaked onto nutrient agar (NA; catalog number CM145; Culture Media and Supplies). All plates were incubated for ~18 h at 37°C and stored at 4°C for no longer than 10 days. Cells from plate cultures were used to inoculate liquid cultures consisting of 10 mL of tryptic soy broth (TSB; catalog number CM104; Culture Media and Supplies) for B. subtilis, B. thuringiensis, S. aureus, E. faecalis, P. aeruginosa, and nutrient broth (NB; catalog number CM146; Culture Media and Supplies) for B. thailandensis. All liquid cultures were incubated for ~18 h at 37°C with rotary aeration at 180 rpm. After incubation, bacteria from liquid cultures were harvested by centrifugation (2300 relative centrifugal force [RCF] at 4°C for 10 min), washed, and resuspended in an equal volume of PBS. The Bacillus species were observed under a microscope to consist predominately of spores. Samples provided for analysis consisted of either a single bacterium or multiple bacteria mixed together. For mixed samples, all bacteria were added in a ratio of 1:1 by volume. All bacteria were present at a concentration between 10E7 and 10E9 cfu/mL as determined by serial dilution and plating onto appropriate agar. All samples were produced at the microbiology laboratory at the U.S. Army ECBC in a blind format and were assigned number codes for processing and analysis. The identities of all blind samples were revealed upon the completion of all analyses. A negative control sample was also included that consisted of PBS only (no bacteria).

7.2.2 PROCESSING OF BLIND BIOLOGICAL SAMPLES

The lysis of all blind samples was performed using a modified sonication method (Ryzhov et al. 2000; Brown et al. 2006; Saikaly et al. 2007). All blind samples, including any sporulated bacteria, were lysed by microprobe ultrasonication (Branson 450 digital sonifier; Branson, Danbury, CT). The blind samples were placed on ice and lysed with a 20 s pulse on and 5 s pulse off (cooling time) and 25% amplitude for a 5 min duration. To verify that the cells were disrupted, a small portion of the lysate was examined under confocal microscopy, and another portion was reserved for 1D gel analysis.

The lysate was centrifuged at 14,100g for 30 min to remove all cellular debris. The supernatant then was added to a Microcon YM-3 filter unit (catalog number 42404; Millipore) and centrifuged at 14,100g for 30 min. The effluent was discarded. The filter membrane was washed with 100 mM ABC and centrifuged for 15–20 min at 14,100g. Cellular proteins were denatured by adding 8 M urea and 3 µg/µl DTT to the filter and incubating it overnight at 37°C on an orbital shaker set to 60 rpm. Twenty microliters of 100% acetonitrile was added to the tubes and allowed to incubate at room temperature for 5 min. The tubes were then centrifuged at 14,100g for 30–40 min and washed three times using 150 I-l of 100 mM ABC solution. On the last wash, the ABC solution was shaken for 20 min, followed by centrifugation at 14,100g for 30–40 min. The filter unit was then transferred to a new receptor tube, and proteins were digested with 5 I-l of trypsin in 240 I-l of ABC solution plus 5 I-l ACN. Protein digestion occurred overnight at 37°C on an orbital shaker set to 55 rpm. Sixty microliters of 5% ACN–0.5% FA was added to each filter to quench the trypsin digestion, followed by 2 min of vortexing for sample mixing. The tubes were centrifuged for 20–30 min at 14,100g. An additional 60 I-l 5% ACN–0.5% FA mixture was added to the filter and centrifuged. Alternative protocols were used in which the denaturation step was eliminated, and the digestion time was reduced using various amounts of trypsin and different digestion temperatures. The effluent was then analyzed using LC-MS/MS.

7.2.3 LC-MS/MS ANALYSIS OF PEPTIDES

The tryptic peptides were separated using a capillary Hypersil C18 column (300 Å; 5 µm; 0.1 mm [inner diameter] by 100 mm) by using the Surveyor LC from Thermo Fisher (San Jose, CA). The elution was performed using a linear gradient from 98% A (0.1% FA in water) and 2% B (0.1% FA in ACN) to 60% B for 60 min at a flow rate of 200 I-l/min, followed by 20 min of isocratic elution. The resolved peptides were electrosprayed into a linear ion trap MS (LTQ; Thermo Scientific) at a flow rate of 0.8 I-l/min. Product ion mass spectra were obtained in the data-dependent acquisition mode that consisted of a survey scan across the m/z range of 400–2000, followed by seven scans on the most intense precursor ions activated for 30 ms by an excitation energy level of 35%. A dynamic exclusion was activated for 3 min after the first MS/MS spectrum acquisition for a given ion. Un-interpreted product ion mass spectra were searched against a microbial database with TurboSEQUEST (BioWorks® 3.1; Thermo Scientific, San Jose, CA) followed by the application of an in-house proteomic algorithm for bacterial identification.

7.2.3.1 Protein Database and Database Search Engine

A protein database was constructed in a FASTA format using the annotated bacterial proteome sequences derived from the sequenced chromosomes of 881 bacteria, including their sequenced plasmids (as of May 2009). A PERL program (ActiveState) was written to automatically download these sequences from the National Institutes of Health NCBI site (http://www.ncbi.nlm.nih.gov). Each database protein sequence was supplemented with information about the source organism and the genomic position of the respective ORF embedded into a header line. The database of bacterial proteomes was constructed by translating putative protein-coding genes and consists of tens of millions of amino acid sequences of potential tryptic peptides obtained by the in silico digestion of all proteins (assuming up to two missed cleavages).

The experimental product ion mass spectral data of bacterial peptides were searched using the SEQUEST (Eng et al. 1994) algorithm against a constructed proteome database of microorganisms. The SEQUEST thresholds for searching the product ion mass spectra of peptides were Xcorr, ΔCn (DelCn), Sp, RSp, and ΔMpep (DelM). These parameters provided a uniform matching score for all candidate peptides. The generated outfiles of these candidate peptides were then validated using the PeptideProphet algorithm (Keller et al. 2002). Peptide sequences with a probability score of 95% and higher were retained in the data set and used to generate a binary matrix of STB assignments. The binary matrix assignment was populated by matching the peptides with corresponding proteins in the database and assigning them a score of one. A score of zero was assigned for a nonmatch. The column in the binary matrix represents the proteome of a given bacterium, and each row represents a tryptic peptide sequence from the LC-MS/MS analysis. Microorganisms in a blind sample were matched with the bacterium/bacteria based on the number of unique peptides that remained after the filtering of degenerate peptides from the binary matrix. The verification of the classification and identification of candidate microorganisms was performed through hierarchical clustering analysis and taxonomic classification (Figure 7.1).

7.2.4 Data Analysis and Algorithms

The SEQUEST-processed product ion mass spectra of the peptide ions were compared to an NCBI protein database using the in house–developed software (ABOid™). ABOid™ (Deshpande et al. 2011) provided a taxonomically meaningful and easy-to-interpret output. ABOid™ calculates the probabilities that a peptide sequence assignment to a product ion mass spectrum is correct and uses accepted spectrum-to-sequence matches to generate an STB binary matrix of assignments. Validated peptide sequences, either present or absent in various strains (STB matrices), were visualized as assignment bitmaps and analyzed by a ABOid™ module that

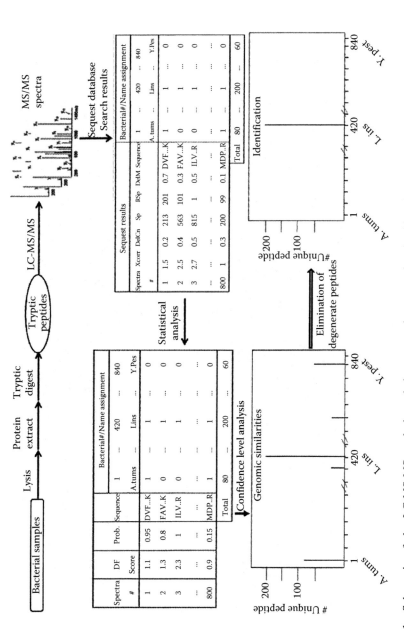

FIGURE 7.1 Schematic of the LC-MS-MS analysis and data-processing steps for the identification of microorganisms using the proteomic approach. *A. tum*, *Agrobacterium tumefaciens*; CID, collision-induced dissociation; df, discriminant factor; *L. inn.*, *Listeria innocua*; *Y. pest*, *Yersinia pestis*. (From Jabbour, R.E. et al., *Appl. Environ. Microbiol.*, 76(11), 3637, 2010.)

used phylogenetic relationships among bacterial species as part of a decision tree process. The bacterial classification and identification algorithm used assignments of organisms to taxonomic groups (phylogenetic classification) based on an organized scheme that begins at the phylum level and follows through the class, order, family, genus, and strain levels. ABOid™ was developed at ECBC using PERL, MATLAB®, and Microsoft Visual Basic, patented (Wick et al. 2012) and licensed to Sage-N Research, Inc., Milpitas, CA, for commercialization.

7.2.5 RESULTS AND DISCUSSION

The capabilities, and possible limitations, of the proteomic approach with regard to the identification of biological agents were evaluated using blind biological samples. Twenty-one blind microbial samples were provided and analyzed by the LC-MS/MS proteomic approach. The composition of the blind samples varied, with some samples having only one bacterium and others having as many as five different bacterial species or strains.

7.2.5.1 *B. subtilis* Sample

An example of the resultant data from the ABOid™ program for one blind sample is shown in Figure 7.2. Blind sample 20 was identified as *B. subtilis* 168 using the ABOid™ data-processing algorithm. This identification algorithm eliminated all of the unwanted and degenerated peptides and retained only the unique peptides that represent a 99% probability for correct identification. In this case, 212 unique peptides were identified and associated with proteins from the *B. subtilis* 168 strain. The 212 *B. subtilis* 168 unique peptides represented 89% of the total number of unique peptides in the blind sample. Table 7.1 shows a selected set of unique peptides and

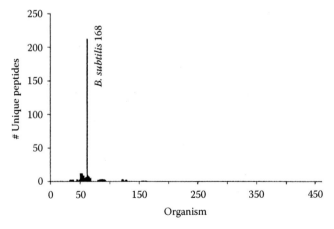

FIGURE 7.2 Histogram of the ABOid™ output for the processing of the LC-MS/MS analysis data set for blind bacterial sample 20. The ordinate provides the actual number of SEQUEST-generated and -filtered unique peptides. The abscissa represents the bacteria found at least once in the 21 experimental samples. (From Jabbour, R.E. et al., *Appl. Environ. Microbiol.*, 76(11), 3637, 2010.)

TABLE 7.1
Partial List of Peptides in the Double-Blind Proteomic Processing of Sample 20, *B. subtilis* 168

Peptide No.	Peptide Sequence	Protein	Accession No.
30	VLDVNENEER	30S ribosomal protein S1	NP_390169.1
39	AYDVSEAVALVK	50S ribosomal protein L1	NP_387984.1
38	NVAVTSTMGPGVK	50S ribosomal protein L1	NP_387984.1
31	GLNVSEVTELR	50S ribosomal protein L10	NP_387985.1
53	GLNVSEVTELRK	50S ribosomal protein L10	NP_387985.1
41	TTPMANASTIER	50S ribosomal protein L13	NP_388030.1
85	GVEMDAYEVGQEVK	50S ribosomal protein L3	NP_387997.1
51	VESPDQLADVLR	Alpha-acetolactate synthase	NP_391482.1
71	EMADFFEETVQK	Aspartyl/glutamyl-tRNA amidotransferase	NP_388551.2
37	EAQQLIEEQR	ATP synthase (subunit b)	NP_391566.1
70	ENTTIVEGAGETDK	Chaperonin GroEL	NP_388484.1
74	AILVMPDTMSMER	Cysteine synthetase A	NP_387954.1
50	LADENSADVYLK	Cysteine synthetase A	NP_387954.1
49	ALSLNETDGFMK	Dihydrolipoamide dehydrogenase	NP_389344.1
64	IGADFLYSVGTLR	Elongation factor EF-2	NP_387993.1
72	SEHGLLFGMPIGVK	Glutamyl-tRNA amidotransferase subunit A	NP_388550.1
100	GILGYSEEPLVSGDYNGNK	Glyceraldehyde-3-phosphate dehydrogenase	NP_391274.1
99	NSSTIDALSTMVMEGSMVK	Glyceraldehyde-3-phosphate dehydrogenase	NP_391274.1
47	TIEVSAERDPAK	Glyceraldehyde-3-phosphate dehydrogenase	NP_391274.1
92	VISWYDNESGYSNR	Glyceraldehyde-3-phosphate dehydrogenase	NP_391274.1
55	TNPDYLFVIDR	Hypothetical protein BSU03830	NP_388265.1
91	AAGATDIYAVELSPER	Hypothetical protein BSU06240	NP_388505.1
35	DIFPAVLSLMK	Hypothetical protein BSU06240	NP_388505.1

(continued)

TABLE 7.1 (continued)
Partial List of Peptides in the Double-Blind Proteomic Processing of Sample 20, *B. subtilis* 168

Peptide No.	Peptide Sequence	Protein	Accession No.
44	GAEIHPNDIVIK	Hypothetical protein BSU06240	NP_388505.1
75	IEHIEEPKTEPGK	Hypothetical protein BSU06240	NP_388505.1
89	EMGHTELPFYQQR	Hypothetical protein BSU12410	NP_389123.1
73	QEETETDLNVLAK	Hypothetical protein BSU12410	NP_389123.1
54	GELEGINFGESAK	Hypothetical protein BSU14180	NP_389301.1
66	SIGVSNFSLEQLK	Inositol utilization protein S	NP_391857.1
79	LISFLQNELNVNK	Isocitrate dehydrogenase	NP_390791.1
5	AVAEALAEAK	Phosphoglycerate kinase	NP_391273.1
94	AVSNPDRPFTAIIGGAK	Phosphoglycerate kinase	NP_391273.1
36	AIQISNTFTNK	Phosphoglyceromutase	NP_391271.1
26	NETVGNAVALAK	Phosphoglyceromutase	NP_391271.1
48	TASVINPAIAFGR	Phosphotransferase system (PTS) fructose-specific enzyme	NP_389323.1
87	IANFETAEPLYYR	Putative manganese-dependent inorganic pyrophosphatase	NP_391935.1
60	LFANLLETAGATR	Ribose phosphate pyrophosphokinase	NP_387932.1
29	TYAQNVISNAK	Serine hydroxymethyltransferase	NP_391571.1
81	FWLSQDKEELLK	S-Ribosylhomocysteinase	NP_390945.1
24	GGPVTLVGQEVK	Thiol peroxidase	NP_390827.1
63	TLGEAVSFVEEVK	Triose phosphate isomerase	NP_391272.1
11	TNDLVADQVK	Triose phosphate isomerase	NP_391272.1

their corresponding proteins that are associated with *B. subtilis* 168. These bacterial proteins have different cellular functions, such as transcription, translation, and cellular signaling. They represent a set of unique biomarkers that could be utilized to establish strain-level discrimination between *B. subtilis* 168 and other members of the *Bacillus* genus.

To ensure confidence in the assignment of the candidate bacterium, a similarity analysis was performed on the nearest-neighbor species and strains. In this similarity analysis, all sequenced strains of *B. subtilis* and *Bacillus* species that are genetically related to the candidate bacterium were included in the Euclidean distance dendrogram. Figure 7.3 shows a dendrogram of the similarity analysis of the blind sample identified as *B. subtilis* 168. In Figure 7.3, the sample was identified as being most similar to *B. subtilis* 168 using the unique peptides that were associated with this bacterial candidate. The next closest bacterium to the candidate was determined to be *Bacillus licheniformis* ATCC 14580. According to these similarities, a comparison of *B. licheniformis* and *B. subtilis* 168 analyses showed a difference of almost 50% in the unique proteins identified by the ABOid™ algorithm. Based on these significant differences and a lower degree of confidence assigned, *B. licheniformis* was not included as a candidate bacterium. Therefore, the identity of sample 20 was assigned to *B. subtilis* 168 using the ABOid™ algorithm. This assignment was correct as latter revealed at the completion of the tests.

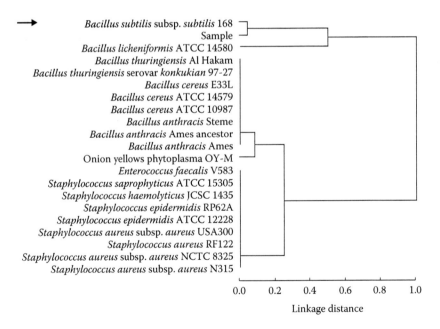

FIGURE 7.3 Dendrogram of the multivariate cluster analysis using Euclidean distances of the STB assignment matrices for blind sample 20. The dendrogram is the result of the complete linkage construction in multivariate data space of the furthest neighbor with the 221 unique peptide sequences shown in Figure 7.2. (From Jabbour, R.E. et al., *Appl. Environ. Microbiol.*, 76(11), 3637, 2010.)

7.2.6   Blind Mixture Analysis

The ABOid™ analysis of sample 18 is shown in Figure 7.4. ABOid™ eliminated all of the unwanted and degenerated peptides, and only the unique peptides that represented a 99% confidence level and above were retained for each organism. In this case, the number of unique peptides varied for the different bacterial candidates. *E. faecalis* had the highest number of unique peptides followed by *B. thuringiensis*, and *B. thailandensis* had the least number of unique peptides. Interestingly, it was revealed that after the tests the blind samples had approximately equivalent bacterial concentrations for each organism, yet the number of unique peptides differed. This variation in the number of unique peptides in the output of the ABOid™ could be due to the dynamic nature of the bacterial species during sample processing. Some bacteria could have a larger number of lysed proteins that were suspended in the extraction buffer than that of other species in the sample. This difference in bacterial protein concentrations is shown in the histogram in Figure 7.4 generated from the ABOid™ output, where the relative number of peptides for each species is compared to that of the other species. This feature in the ABOid™ algorithm could be used as a pseudo-quantitative technique in the determination of lysed bacterial proteins in a biological sample and thus aided in evaluating sample-processing modules. Also shown in Figure 7.4 are six bacterial candidates near the cutoff threshold within the *Staphylococcus* genus. This pattern is due to the fact that the *S. aureus* ATCC 3359 strain present in the blind sample has not been sequenced or reported in the public domains, and thus it was not part of the constructed proteome database. However, the ABOid™ was capable of providing a nearest-neighbor match to the species level (*S. aureus*) and thus identified the bacterium correctly as *S. aureus* subsp. *aureus*.

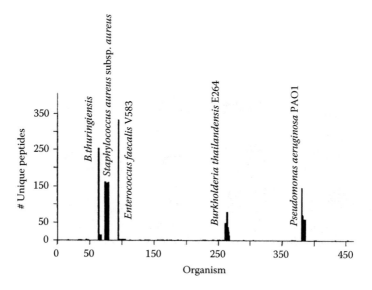

FIGURE 7.4   Histogram of the ABOid™ BACid output for the processing of the LC-MS/MS data set for the biological mixture in sample 18. Refer to the legend of Figure 7.2 for details. (From Jabbour, R.E. et al., *Appl. Environ. Microbiol.*, 76(11), 3637, 2010.)

It is noteworthy that this bacterial strain, which is not genomically sequenced, could be identified only to the species level. The rapid increase in the number of sequenced bacteria will benefit this proteomic approach and enhance its robustness in the identification process of biological samples. However, a significant advantage of the approach is that if a particular strain has not been sequenced but the species is represented in the database, it is highly likely that the unsequenced sample strain will be identified to the species level. The appearance of the histogram from a ABOid™ analysis indicates the degree of the accuracy of the identification process. Strain-level experimental identification is indicated by a single line (Figure 7.4) in the histogram (*E. faecalis* V538) or by a grouping of lines, where one line clearly dominates (e.g., *B. thailandensis* E264 and *P. aeruginosa* PAO1) with respect to the number of unique peptides. *B. thuringiensis* has two strains resident in the database, and both provide a similar set of peptides. This occurs because the two strains do not display peptides that clearly distinguish themselves. The fifth bacterium in the sample 18 mixture was *S. aureus* strain ATCC 3359, and this organism's genome has not been sequenced. However, the species-level identification (*S. aureus*) of this strain is indicated by a grouping of lines (Figure 7.4) that does not display a significant difference in the number of unique peptides. This blind sample was correctly identified as a mixture of five bacteria: *B. thuringiensis*, *S. aureus* subsp. *aureus*, *E. faecalis* V583, *B. thailandensis* E264, and *P. aeruginosa* PAO1, where *S. aureus* and *B. thuringiensis* were identified to the species level and the other three were identified to the strain level.

7.2.7 SUITE OF BACTERIAL SAMPLES

The in-house database originated from 881 genomically sequenced bacterial strains. The blind sample suspensions consisted of bacteria in single and mixture forms, and their genomes were sequenced or not sequenced. The bacterial strains found in experimental samples that do not have a sequenced genome, therefore, cannot be found in available public databases and the in-house database. Figure 7.5a shows the classification map of the 21 experimentally processed blind samples, and Figure 7.5b shows that of the bacterial strain sample identities (sample key). In Figure 7.5a, the bacteria on the abscissa reflect every bacterium found at least once in the 21 experimentally determined samples. The bacteria listed in Figure 7.5a were not disclosed in advance; rather, all 21 experiments produced the bacterial identities from the ABOid™ algorithm (Deshpande et al. 2011). Figure 7.5b represents the sample key or actual bacterial species and strains in the blind samples. This information was not released to the authors until Figure 7.5a results were turned in for experimental performance verification. A comparison between Figure 7.5a and b shows that bacterial discrimination was achieved by relying on the unique peptides corresponding to the bacteria in the blind samples. An identification was based on the matching probability of the unique peptides from a blind sample with a bacterial entry in the bacterial proteome database at more than a $p = 0.95$ confidence level. The strain-level identification, indicated by the filled red boxes in Figure 7.5a, was assigned due to a close match with the analyzed microorganisms' unique peptides and their nearest-neighbor strains.

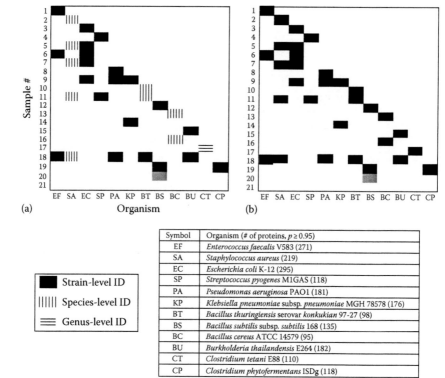

Symbol	Organism (# of proteins, $p \geq 0.95$)
EF	*Enterococcus faecalis* V583 (271)
SA	*Staphylococcus aureus* (219)
EC	*Escherichia coli* K-12 (295)
SP	*Streptococcus pyogenes* M1GAS (118)
PA	*Pseudomonas aeruginosa* PAO1 (181)
KP	*Klebsiella pneumoniae* subsp. *pneumoniae* MGH 78578 (176)
BT	*Bacillus thuringiensis* serovar *konkukian* 97-27 (98)
BS	*Bacillus subtilis* subsp. *subtilis* 168 (135)
BC	*Bacillus cereus* ATCC 14579 (95)
BU	*Burkholderia thailandensis* E264 (182)
CT	*Clostridium tetani* E88 (110)
CP	*Clostridium phytofermentans* ISDg (118)

■ Strain-level ID

||||| Species-level ID

≡ Genus-level ID

FIGURE 7.5 (a) Classification map of the experimentally processed samples. The bacteria on the abscissa indicate that they were found at least once in the 21 samples; (b) actual or sample key of bacteria present in all 21 samples. The dark gray coloring for sample 20 represents *B. atrophaeus*, which was identified as *B. subtilis* in panel (a). Sample 21 was a blank. In the table, the numbers in parentheses for each organism signify the number of proteins identified with a $p > 0.95$ probability match. Solid box, strain-level identification; vertically hatched box, species-level identification; horizontally hatched box, genus-level identification. (From Jabbour, R.E. et al., *Appl. Environ. Microbiol.*, 76(11), 3637, 2010.)

Figure 7.4 shows the analysis of sample 18 and provides an example of identification to the strain level as well as classification to the species level (as described earlier) for *S. aureus* strain ATCC 3359, which is not currently sequenced. A correct species level of identification was experienced with all bacteria in the blind samples that are unsequenced and are indicated by a vertical hashed box in Figure 7.5a. Thus, the classification probability was statistically high enough based on a comparison of the virtual proteome of a database strain and the experimental unique proteins of the unsequenced genome bacterial sample. Therefore, identification was reported at the species level. Blind sample 20 (Figure 7.2) was identified as *B. subtilis*; however, the sample key reported it as *B. atrophaeus*. This difference is due to the lack of a proteome for *B. atrophaeus*, which taxonomically is considered *B. subtilis*. Our data support the proposition that *B. atrophaeus* should be reclassified as a strain of *B. subtilis* (Burke et al. 2004) (the gray square for sample 20 in Figure 7.5a and b).

7.2.8 GENUS-LEVEL IDENTIFICATION

Blind sample 17 was investigated for ABOid™ characterization. The experimental set of peptides could provide results only to the *Clostridium* genus level, because all nine clostridial bacteria (species and strains) resident in the database produced a histogram (data not shown) similar to that of *S. aureus*, which is shown in Figure 7.4. The experimental peptides matched that portion of the virtual proteome common to all clostridia. Therefore, the complete experimentally derived tryptic peptide information record was stored as a separate bacterial line item as *Clostridium* species 1 in the database of 881 bacteria. Another aliquot of the blind sample was processed with data reduction and searching in the new hybrid database. The highest match was with the *Clostridium* species 1 entry. After the results were submitted, the identity of sample 17 was revealed to be *C. phytofermentans* ISDg. This strain does not have its genome sequenced, yet ABOid™ was able to match the virtual proteins that are similar to those of the *Clostridium* genus to the experimentally observed peptides. Thus, ABOid™ was able to characterize sample 17 as *Clostridium* without choosing one of the nine clostridia strains resident in the database or other bacterial genera. ABOid™ instead matched *Clostridium* species 1 to the experimental peptides, which indicated that there is sufficient information in the experimental peptides to differentiate *C. phytofermentans* ISDg from the nine database clostridia strains. It is tempting to consider that this approach, when combined with the accurate mass tag approach of Lipton et al. (2002), has the potential to diminish the impact of genome-sequencing deficiencies for some bacterial strains. The rapid advancement in genome-sequencing projects will enhance the robustness of this approach through the expansion of the proteome database. This expansion in the proteome database is anticipated to include the cellular proteins that can be utilized for strain-level differentiation.

The results showed that the method was effective in identifying bacteria whether the sample was composed of one organism or a mixture, or even if the sample is not resident in the database. No false positives were observed for any of the blind samples that were analyzed, including the blank sample. The proteomic MS approach reported herein is not meant as a replacement for DNA-based identification methods. We envision this approach as a second, confirmatory approach to pathogen identification. Additionally, there are some major advantages to the proteomic method over other molecular biology methods such as the DNA-based methods, in that (i) no prior information about the sample is required for analysis; (ii) no specific reagents are needed in the analysis process; (iii) proteomic MS is capable of identifying an organism when a primer/probe set is not available; (iv) proteomic MS requires less rigorous sample preparation than PCR; and (v) proteomic MS can provide a presumptive identification of a true unknown organism by mapping its phylogenetic relationship with other known pathogens. The proteomic method could also be applied to identify viruses and toxins, because viruses and toxins are included in the proteome database.

Naturally occurring environmental samples usually contain a great deal of organisms at very low concentrations in addition to the target species. The total amount of background organisms may consist of greater numbers than that of the target organism. Therefore, this is a topic that would challenge the method reported herein. This is being addressed by spiking a target organism in several environmental matrices at different applied amounts.

Improvement in sample preparation and MS technologies will enhance and increase the number of peptides identified compared to those of the current methods. This can allow for MS proteomics being a valuable tool in conjunction with genomic approaches to address the issue of the identification and classification of microorganisms. Overall, these studies showed that the MS-based proteomic approach is a useful method that may be applied to diverse biothreat scenarios and has the potential for bacterial differentiation and identification at species and strain levels of individual bacteria or their mixtures.

7.2.9 Comments

The methods used in this double-blind study were applied to many different types of samples including food, human fluids, animals, plants, and general environmental sources. Bacteria, viruses, and fungi were detected accurately and repeatedly. One interesting observation was made when examining mashed potato samples from a food source. The bacteria of question were detected, but also the thermophiles that were in the hot water used to prepare them. These thermophiles were considered to be part of the microbial signature of these samples as other sources of the mashed potatoes did not contain them. Likewise, many other microbes were frequently detected in samples and these represented the more complete analysis.

7.3 PANDEMIC H1N1 2009 VIRUS

The Pandemic H1N1 2009 virus caused major concerns around the world because of its epidemic potential, rapid dissemination, rate of mutations, and the number of fatalities. One way to gain an advantage over this virus is to use existing rapid bioinformatics tools to examine easily and inexpensively generated genetic sequencing data. We have used the protein sequences deposited with the NCBI for data mining to study the relationship among the Pandemic H1N1 2009 proteins. There are 11 proteins in the Pandemic H1N1 2009 virus, and analysis of sequences from 65 different locations around the globe has resulted in two major clusters. These clusters illustrate that the Pandemic H1N1 2009 virus is already experiencing significant genetic drift and that rapid worldwide travel is affecting the distribution of genetically distinct isolates.

The H1N1 strain of influenza is a single-stranded RNA virus composed of a segmented genome originated from various influenza viruses (Solovyov et al. 2009). An infection of mixtures of various influenza viruses results in the release of progeny viruses containing novel arrangements of segments. In Asia, North America, and much of Europe, viruses of the H1N1 subtype are the most commonly isolated (Campitelli et al. 1997; Scholtissek et al. 1998). However, for the purposes of this study, we have chosen to focus on the Pandemic H1N1 2009 virus isolates (Fraser et al. 2009) that have been of great worldwide public concern. The Pandemic H1N1 2009 viruses differ in the origins of their genomic components from these previously circulating H1N1 strains and belong to the classic swine lineage, which is genetically related to the human H1N1 viruses responsible for the 1918 Spanish influenza pandemic (Schultz et al. 1991; Reid et al. 1999). The reassortment and selection events that led to the recent outbreak of influenza viral infection in man could be traced to several factors including cellular receptors, oligosaccharides, the cohabitation of swine and poultry, and the presence of avian and human receptors in pigs (Done and Brown 1999; World Health Organization 2010).

The first reported outbreak of the mutated subtype influenza virus, Pandemic H1N1 2009 occurred in Mexico in April 2009 and was rapidly followed by more cases reported in the United States and many other countries. According to the World Health Organization (2010), in a 2 month period, the Pandemic H1N1 2009 virus spread to more than 65 countries and infected more than 17,000 people, with 115 deaths.

Several studies have recently been reported that addressed the origin of the Pandemic H1N1 2009 using cluster analyses of genomic segments isolated from various infection cases. The overall agreement traces the Pandemic H1N1 2009 to a "triple-assortment" swine influenza predominant in North America and H1N1 strains predominant in swine populations in Europe and Asia (Novel Swine-Origin Influenza [H1N1] Investigation Team 2009; Solovyov et al. 2009). While such studies are of significant importance to the public and scientific communities, the correlation studies among the worldwide reported cases of Pandemic H1N1 2009 would expand the scope of coverage for this epidemic. This paper addresses the Pandemic H1N1 2009 protein sequences collected from NCBI (2009) as of June 11, 2009.

These sequences were specific to 2009 outbreak only, and originated from 65 different global locations (National Center for Biotechnology Information 2010). Our results show that there is significant clustering of Pandemic H1N1 2009 among the cases. Also, some cases with reported antiviral resistance (oseltamivir) showed the same mutation of a single amino acid H274Y in the neuraminidase (NA) protein of H1N1 that are in agreement with relevant reported data.

7.3.1 Materials and Methods

The Pandemic H1N1 2009 protein sequences were downloaded from the NCBI web site (National Center for Biotechnology Information 2009) in FASTA format as of June 11, 2009. These protein sequences were specific to the 2009 outbreak, and were collected from different infected people from 65 different locations (National Center for Biotechnology Information 2010).

The combined FASTA file was fragmented into location-specific protein sequences. These location-specific FASTA files were then uploaded into a Structured Query Language (SQL) Server relational database management system (RDBMS). This database was used to create a 65×65 correlational matrix based on the locations. The corresponding matrix FASTA files were merged to create a two location-based protein sequence file; a total of 4225 FASTA files were created in this matrix. These files were indexed using DBIndexer3 utility included in its BioWorks Suite of applications (Thermo Fisher). The generated header files were used to read the unique number of peptides observed and the computed values were assigned in the corresponding cell of the matrix.

Jaccard's index was computed for each cell to determine the similarity of a given cell to the data set. The unique peptide values observed for each specific location were used to perform clustering calculations, using the linkage rule of "Wards Method" and distance measure of "Euclidean distance." This analysis was automated in-house with coded software application labeled as "genTree."

7.3.2 RESULTS AND DISCUSSIONS

The resulting cluster analysis generated from the protein sequences of the Pandemic H1N1 2009 virus showed two distinct sub-clusters (Figure 7.6). The distinct sub-clusters vary in their number of the locations with cluster A containing seven countries, and sub-cluster B 17 countries (Table 7.2). Also, sub-cluster A has fewer numbers of cases (29) than that of sub-cluster B (36), although this may be the result of availability of sequence data rather than an indication of a true disparity in total numbers of cases. The proteomic variability of the cases within a sub-cluster shows some independence of the geographical location, which may demonstrate dissemination of distinct genetic isolates via commercial airline travel. For example in the

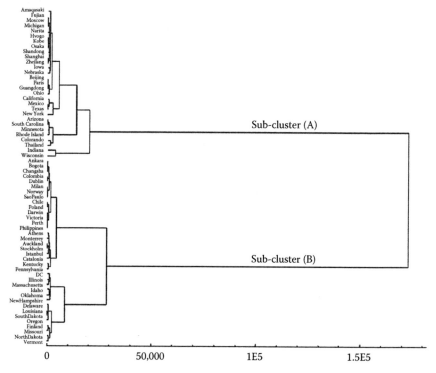

FIGURE 7.6 Cluster analysis of Pandemic H1N1 2009 reported cases from different locations. The horizontal length represents the similarity among the protein sequences, and the vertical lines indicate spacing. (From Wick, C.H. et al., 2009 Cluster analysis: A preliminary assessment, *Nature Precedings*, September 16, 2009, http://precedings.nature.com/documents/3773/version/1/files/npre20093773–1.pdf, accessed July 1, 2012.)

TABLE 7.2
Country Listing by Sub-Cluster

Sub-Cluster A		Sub-Cluster B	
China	Australia	Italy	Sweden
France	Chile	Mexico	Turkey
Japan	China	New Zealand	United States
Mexico	Columbia	Norway	
Russia	Finland	Philippines	
Thailand	Greece	Poland	
United States	Ireland	Spain	

---

topmost portion of sub-cluster A, a significant grouping was observed among cases from Japan (6 cases), China (3 cases), United States (1 case), and Russia (1 case). This variability pattern is also observed with other cases in either sub-cluster. While it is expected to have same location cases grouped together, the presence of cases from geographically distant locations indicates other factors, such as airline travel, that should be considered.

The two sub-clusters were also characterized by variation in the serotype of the Pandemic H1N1 2009 strains. Most notable are cases from China and United States which have the largest variability, as cases from these two locations are spread across both sub-clusters, perhaps an indication of individual mobility. The serotypes of the Pandemic H1N1 2009 infecting humans in California are similar to those found in Texas and Mexico (sub-cluster A), but different from those found in Vermont, DC, and Pennsylvania cases (sub-cluster B). Overall, this clustering analysis potentially provides a different perspective on the Pandemic H1N1 2009 strains in terms of geographic distribution of population and migration factors. The establishment of two distinct clusters from the cases studied is an indication of strain variability of Pandemic H1N1 2009 to be considered for diagnosis and prognosis purposes. This kind of automated bioinformatics analysis may be useful for the ongoing assessment of how viruses such as influenza spread across the globe, monitoring spread in drug-resistant or particularly virulent genotypes, and potential antigenic/genetic shifts that may impact detection and vaccine efficacy. The addition of more sequence information, easily and cheaply available, correlated with clinical and phenotypic data would be of particular interest to the public health community.

7.3.3 COMMENTS

Using the same methods, single changes in the N protein were detected for two strains of the virus that were resistant to antiviral protocols. Comparing sequenced microbes in this manner is quick and provides a means for following changes in the genome over the epidemic. This process can be followed for other microbes and other applications. It is an excellent tool for following emerging microbes of all types.

7.4 ANALYSIS OF HONEYBEES

7.4.1 INTRODUCTION AND BACKGROUND

In 2010, colony collapse disorder (CCD) again devastated honeybee colonies in the United States, indicating that the problem neither is diminishing nor has it been resolved. Many CCD investigations, using sensitive genome-based methods, have found small RNA bee viruses and the microsporidia, *Nosema apis* and *Nosema ceranae*, in healthy and collapsing colonies alike with no single pathogen firmly linked to honeybee losses.

7.4.1.1 Methodology/Principal Findings

We (Bromenshenk et al. 2010) used MSP to identify and quantify thousands of proteins from healthy and collapsing bee colonies. MSP revealed two unreported RNA viruses in North American honeybees, Varroa destructor-1 virus and Kakugo virus, and identified an invertebrate iridescent virus (IIV) (Iridoviridae) associated with CCD colonies. Prevalence of IIV significantly discriminated among strong, failing, and collapsed colonies. In addition, bees in failing colonies contained not only IIV, but also *Nosema*. Co-occurrence of these microbes consistently marked CCD in (1) bees from commercial apiaries sampled across the United States in 2006–2007, (2) bees sequentially sampled as the disorder progressed in an observation hive colony in 2008, and (3) bees from a recurrence of CCD in Florida in 2009. The pathogen pairing was not observed in samples from colonies with no history of CCD, namely bees from Australia and a large, non-migratory beekeeping business in Montana. Laboratory cage trials with a strain of IIV type 6 and *N. ceranae* confirmed that coinfection with these two pathogens was more lethal to bees than either pathogen alone.

7.4.1.2 Conclusions/Significance

These findings implicate coinfection by IIV and *Nosema* with honeybee colony declines, giving credence to older research pointing to IIV, interacting with *Nosema* and mites, as probable cause of bee losses in the United States, Europe, and Asia. We next need to characterize the IIV and *Nosema* that we detected and develop management practices to reduce honeybee losses. History of CCD is described elsewhere.

7.4.2 MSP

MSP and a rigorous sampling method were used in an attempt to identify potential markers of CCD. MSP offered an orthogonal and complementary approach (Aebersold and Mann 2003; Dworzanski et al. 2006) to gene-based techniques used in previous CCD studies for pathogen screening and classification. MS yielded unambiguous peptide fragment data that were processed by bioinformatics tools against the full library of peptide sequences based on both genomic and proteomic research.

Consequently, peptide fragment data acquired by MSP allowed identification and classification of microorganisms from the environment that was unrestricted by the need for amplification, probes, or primers. Furthermore, this approach allowed for the detection, quantification, and classification of fungi, bacteria, and viruses in a single analytical pass (Keller et al. 2002; Jabbour et al. 2010b; Deshpande et al. 2011). Classification can be to strain level and is limited only by the level of precision within the proteomic and genomic databases.

Our MSP analyses revealed the presence of two RNA viruses not previously reported in North American bee populations, as well as a highly significant and also unreported co-occurrence of strains of DNA IIV with a microsporidian of the genus *Nosema* in CCD colonies. The two RNA viruses were only seen occasionally, but the finding of the DNA virus in virtually all CCD samples establishes a new avenue for CCD research, as nearly all previous viral work to date in honeybees has focused on RNA viruses.

7.4.3 RESULTS

MSP analysis was used to survey microbes in bee samples from (1) CCD colonies from the original event in 2006–2007 from widespread locations in eastern and western parts of the United States (2006–2007 CCD colonies), (2) a collapsing colony in an observation hive fitted with a bidirectional flight counter and sampled through time as it failed in 2008 (2008 Observation Colony), (3) an independent collapse of bee colonies from CCD in Florida in 2009 (2009 Florida CCD), (4) packages of Australian honeybees delivered to the United States (2007 Australian Reference Group), (5) an isolated non-migratory beekeeping operation in Montana with no history of CCD (Montana Reference Group), and (6) dead bees recovered from inoculation feeding trials with *N. ceranae* alone, IIV alone, a mixture comprising *N. ceranae* plus IIV, and controls that were fed syrup alone in 2009–2010 (Inoculations Recovery Group).

MSP analysis resulted in a database of more than 3000 identifiable peptides, representing more than 900 different species of invertebrate-associated microbes. An extensive summary of detected peptides and microbes is presented in a recently completed technical report (Wick et al. 2010). We narrowed the list of suspect microbes to those infecting bees and insects, 121 in all. Of these, only 29 were specific to bees or occurred in more than 1% of the colonies sampled. These formed the subset of pathogens that we used for subsequent analyses. We focused our search on viruses, fungi, and microsporidia in the genus *Nosema*. We did not include well-known bacterial infections of honeybees that are easily recognized, with visible signs that differ from CCD. We also observed Varroa mites in some, but not all of the CCD colonies.

Peptides were identified from 9 of the ~20 known honeybee viruses in the initial sample set (Table 7.1). The isolated, non-migratory, Montana colonies that we included as a reference group were unique in that they were nearly virus-free except for a single colony that was positive for a low level of sacbrood virus (SBV) infection.

The recently described Varroa destructor virus 1 (VDV-1) (Ongus et al. 2006) was detected in two colonies from the 2006–2007 CCD colonies; one from an eastern, and one from a western location (Michel 2008; Wick et al. 2010). Peptides of Kakugo virus (Fujiyuki et al. 2006; Terio et al. 2008), which previously has not been reported in North American bees, were detected in two colonies from a single west coast location in this same CCD group of colonies.

Israeli acute paralysis virus (IAPV) did not occur frequently and was distributed approximately equally among strong and failing colonies (Table 7.3). It was more prevalent in colonies that originated from the east coast of the United States (4 of 10) and Australia (3 of 10).

The most prevalent viral peptides we detected were identified as IIV, large double-stranded DNA viruses of the Iridoviridae family. We detected 139 unique peptides in west and east coast colonies that were attributed only to IIV type 6 (IIV-6, also known as Chilo iridescent virus) with high confidence (≥ 0.99).

IIV appeared with 100% frequency and at higher peptide counts in failing and collapsed colonies. IIV also occurred in nearly 75% of strong colonies, although invariably at lower concentrations. Numerous peptides for *Nosema* species were detected in collapsed and failing colonies. Peptides attributed to 10 species of *Nosema* were represented, but because of high cross-correlations among the different peptides within the genus these were aggregated based on cluster analysis into two distinct groupings.

Using those groupings, we observed that one group of *Nosema* peptides paralleled the pattern of occurrence for IIV ($r=0.90$, $n=31$, $p=0.001$) and was present at higher frequency more often in failing and collapsed colonies than in strong colonies (Table 7.3). Further suggestive correlations with other microbes included the co-occurrence of black queen cell virus (BQCV) and IIV ($r=0.71$, $n=31$, $p=0.001$) and concordantly the same *Nosema* group ($r=0.73$, $n=31$, $p=0.001$).

Count-weighted occurrence data were subjected to stepwise discriminate function analysis to assess whether strong, failing, or collapsed colonies could be differentiated by specific patterns of pathogen occurrence. The isolated Montana apiary was included as a non-CCD reference group for this analysis. The colonies in this group served as an external control group that was complementary to the strong colonies within the CCD apiaries that served as internal controls.

Discriminate analysis indicated that only two pathogens, IIV and deformed wing virus (DWV), were necessary for significant discrimination among different colony groups (Table 7.4). The first function contrasted higher incidence of IIV in failing colonies with higher incidence of DWV in the remaining groups (Figure 7.7). The structure matrix (correlations with discriminant functions) showed that *Nosema* 1 was the highest correlated variable among those not included with the discriminant functions (Table 7.5, $r=0.69$). This indicates that the incidence of IIV and *Nosema* 1 was strongly associated with group scores on the discriminant functions.

As expected, the Montana reference group was most distinct and significantly different from the strong condition colonies ($p_{\text{out-strong}}=0.06$, $F=5.5$, $df=2.33$; $p_{\text{out-failing}}=0.001$, $F=17.3$, $df=2.33$; $p_{\text{out-collapsed}}=0.04$, $F=7.5$, $df=2.33$). Failing colonies were distinct from both good and reference colonies ($p_{\text{failing-strong}}=0.002$, $F=7.4$, $df=2.33$; $p_{\text{failing-out}}=0.001$, $F=10.1$, $df=2.33$) based mostly on differences in IIV peptide abundance. The only anomaly was that collapsed and strong colonies were not significantly different ($p_{\text{collapse-strong}}=0.71$, $F=0.3$, $df=2.33$). This similarity between collapsed and strong colonies seems contradictory at first.

It is, however, likely that the few bees left in colonies at the final stages of collapse are those that are not infected, and thus would be expected to be similar to uninfected bees in strong colonies. *Nosema* species by themselves were not a significant

TABLE 7.3

Frequency (Frq) of Occurrence and Mean Peptide Counts of Viral Pathogens and *Nosema* in Honeybee Colonies Sampled in 2006, 2007, and 2008[a]

Pathogen	East Coast–West Coast Colonies, 2006						Observation Colony, 2007		Florida Colonies, 2008	
	Collapsed n=8		Failing n=10		Strong n=13		Sub-Samples=18		n=9	
	Frq	x̄ (s.d.)	Frq	x̄ (s.d.)	Frq	x̄ (s.d.)	Frq	x̄ (s.d.)	Frq	x̄ (s.d.)
Acute bee paralysis virus	2	0.3 (0.46)	5	1.5 (2.07)	5	0.9 (1.28)	13	1.3 (1.28)	7	11.6 (12.4)
Black queen cell virus	2	0.4 (0.74)	6	1.4 (1.8)	3	0.8 (1.54)	4	0.3 (0.57)	7	1.9 (1.5)
Deformed wing virus	3	0.8 (1.4)	1	0.2 (0.6)	6	0.6 (0.8)	4	0.6 (1.38)	7	15.9 (20.1)
Iridescent virus	8	20.9 (28.2)	10	38.0 (39.6)	9	15.6 (22.4)	18	16.1 (12.74)	9	57.6 (23.6)
Israeli acute paralysis virus	1	0.3 (0.7)	4	1.4 (2.3)	5	0.8 (1.3)	11	0.9 (0.96)	5	2.4 (2.8)
Kakugo virus	0	0 (0)	0	0 (0)	3	0.3 (0.08)	3	0.2 (0.55)	2	0.3 (.04)
Kashmir bee virus	3	0.2 (3.2)	6	1.9 (2.1)	9	1.0 (0.9)	1	1.0 (1.28)	6	3.6 (5.0)
Sacbrood virus	2	0.9 (1.6)	4	0.9 (1.4)	6	1.2 (2.3)	11	1.3 (1.36)	6	3.8 (7.0)
Varroa destructor virus 1	0	0 (0)	1	0.2 (0.6)	1	0.2 (0.6)	4	0.4 (1.04)	5	1.3 (1.6)
Nosema group 1	5	6.4 (9.1)	9	11.4 (9.6)	7	5.2 (7.7)	18	8.7 (5.74)	9	35.2 (15.3)
Nosema group 2	3	0.8 (1.4)	3	0.7 (1.3)	3	0.2 (0.4)	11	1.0 (0.97)	0	0 (0)

Source: Bromenshenk, J.J. et al., *PLoS One*, 5(10), e13181, 2010.

[a] Columns summarize 31 colonies from initial CCD study in 2006; 18 sub-samples taken from an observation colony monitored through its collapse from March through August 2007; and a third sample of nine colonies sampled during a CCD incident in Florida in 2008. A hyphen indicates that the value could not be calculated.

Both of these pathogens increased when the bee population decreased most sharply and remained at high levels throughout the remainder of the collapse. None of the other pathogens found in the colony showed a similar pattern. PCR analysis of the *Nosema* species present in this observation colony revealed the presence of *N. ceranae* alone (data not shown). The 2009 Florida CCD samples presented an independent opportunity to corroborate our findings. A group of nine colonies, in sets of three, identified as either strong, failing, or collapsed, were sampled and analyzed. We used the classification functions generated in our original discriminant function analysis from 2007 to classify these new samples as either strong, failing, collapsed, or out-group reference based on virus and *Nosema* patterns. Of the nine colonies, six were classified as either collapsed or failing. None were classified as the healthy reference group, leaving the remaining three classified as strong. The analytical classification was close to the original designation for these colonies. Two of the three originally identified strong colonies were classified as such; one was classified as failing.

7.4.3.1   Differences between Failing and Figure 7.7

Discriminant function analysis for differences in pathogen peptide counts among strong, failing, and collapsed honeybee colonies. Function I explains 81% of discriminating variance and contrasts higher incidence of iridovirus (IIV), *Nosema*, and to a lesser extent BQCV in failing colonies with higher incidence of DWV and some IAPV in the remaining groups. Vertical and horizontal lines mark the non-CCD out-group as a reference set. Collapsed designations were not as distinct, but were consistent in classifying five of six colonies as suffering from CCD based on IIV and *Nosema* occurrence.

We detected 139 unique peptides in our west and east coast data that were attributed to IIV-6 with high confidence (match to index ≥ 0.99). Later samples also indicated an IIV-6-like virus as the dominant virus in collapsing colonies (88% of IIV peptides). Furthermore, comparison of IIV peptides among all samples revealed moderate but significant correspondence between the original samples, the laboratory inoculation experiments, and the other field samples (Table 7.6).

These procedures may have identified IIV-6 as the most likely source of peptides because this is the only fully sequenced genome from the genus *Iridovirus*. We suspect that bees may in fact be infected by IIV-24 that is also assigned to the *Iridovirus* genus, which was isolated from an Asian bee (Bailey et al. 1976; Bailey and Ball 1978; Verma and Phogat 1982) or by a variant of IIV-6. Isolation of the unknown IIV we detected is not a trivial matter. Virus isolation is usually achieved using in vitro cell culture techniques, but this is not easily achieved from CCD bees because many are infected by multiple viruses and plaque purification techniques are not available for most of these viruses. Nonetheless, attempts at viral isolation are ongoing.

Our original wide area bee samples from 2007, the time series samples from the collapsing observation colony from 2008, and the reoccurrence of CCD reflected by Florida samples from 2009 were consistent in that IIV-6 peptides plus some IIV-3 peptides were the only IIV entities detected and were correlated with infections by *Nosema* species. Isolation of the IIV(s) and *Nosema* species

TABLE 7.6

Similarity in Occurrence of Specific Iridescent Virus Peptides among Different Samples Analyzed for Evidence of Pathogens Associated with CCD

Sample		Florida Collapse	Inoculation Trial	Collapsing Colony
Inoculation trial	Rho[a] (p two tailed)	0.26 (0.00)		
	Sorensen's index[b]	0.18		
Collapsing colony	Rho (p two tailed)	0.08 (0.21)	0.11 (0.07)	
	Sorensen's index	0.21	0.18	
East–West CCD colonies	Rho (p two tailed)	0.30 (0.00)	0.22 (0.00)	0.03 (0.66)
	Sorensen's index	0.58	0.20	0.17

[a] Spearman's rank correlation (rho; $n = 266$).

[b] Sorensen's index of similarity was calculated for each pairwise comparison. East–West CCD colonies sampled 2007–2008; Collapsing observation colony, 2008; Florida collapse, 2009; Inoculation trials, 2009–2010.

in our samples is not trivial and is the subject of our ongoing research. Isolates, however, of a strain of IIV-6 and *N. ceranae* were immediately available. Thus, to test our MSP generated hypothesis that an interaction between *N. ceranae* and IIV leads to increased bee mortality we conducted inoculation cage-trial experiments.

7.4.4 Cage Trials

Cage trials of 1–3 day old newly emerged bees demonstrated increased mortality in the experimental group fed both *N. ceranae* and IIV-6 in comparison with the control group ($p = 0.0001$) and bees fed only *N. ceranae* ($p = 0.04$) or only IIV-6 ($p = 0.04$, Figure 7.9). As the actual infectious dose of *N. ceranae* or IIV that occurs in the field is currently unknown, we chose to utilize a relatively low infectious dose for both pathogens in our experiments. As is common in cage bee trials, mortality was observed in the control groups in all four biological replicates. To confirm that the controls likely died from a noninfectious cause, deceased bees from all treatment groups were further screened with MSP. The controls did not have any detectable IIVs, but did show some evidence of *Nosema*, which was not apparent from PCR analysis of the same samples.

These results revealed mostly an absence of pathogens in the control bees, and the presence of peptides related to IIV-6 or *N. ceranae* in the appropriate groups (data not shown). Importantly, no statistical difference in survival was observed between the bees fed *N. ceranae* or the virus alone. These results support the correlation observed by the MSP data that suggests that an interaction between *N. ceranae* and an IIV-6-like virus may be involved in bee mortality. Whether an additive or synergistic interaction would be observed between *N. ceranae* and other bee viruses is currently unknown, but it merits further study.

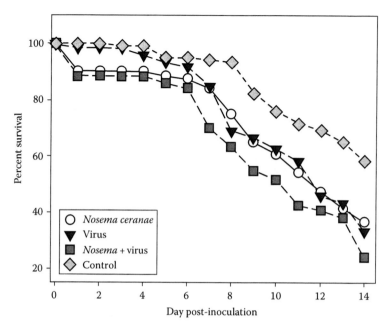

FIGURE 7.9 Survival over a 14 day postinfection period observed in cage trials of *N. ceranae*– and IIV-infected honeybees. Newly emerged, 1–3 day old, bees were used in all experiments. The figure represents the combined survival results for four biological replications ($N = 30$ bees in each group for each biological replicate). Bees that perished within 24 h of inoculation were not included in the survival curve analyses. Deaths in the control group were confirmed not to be pathogen related via MS analyses. Inoculum sizes and doses were described in Section 7.2.1. Log-rank tests Kaplan–Meier curve analyses: (1) control vs. *N. ceranae*, $p = 0.01$, (2) control vs. IIV alone, $p = 0.01$ (0.008), (3) control vs. *Nosema*+IIV, $p = 0.01$ (0.0001), (4) *Nosema* alone vs. IIV alone, $p = 0.90$, (5) *Nosema* alone vs. *Nosema*+IIV, $p = 0.04$, and (6) virus alone vs. *Nosema*+IIV, $p = 0.04$. These results strongly suggest that the combination of *N. ceranae* and IIV is associated with increased bee mortality.

7.4.5 DISCUSSION

MS-based proteomics provided an unrestricted and unbiased approach for surveying pathogens in honeybee colonies. Our results detected a DNA virus and two RNA viruses that had not been previously reported in honeybees from the United States. The potential correlation of IIV with CCD may previously have gone unnoticed because these are large DNA viruses, not the small RNA viruses commonly considered to be the cause of most bee diseases. The correlation between IIV and *N. ceranae* for bee colonies exhibiting CCD implies that they track each other.

Interestingly, the presence or absence of IIV in a given honeybee colony may explain why in the United States *N. ceranae* sometimes seems to contribute to severe colony losses (IIV present), and sometimes not (IIV absent), as reported both by researchers and beekeepers (CCD Working Group 2006; Cox-Foster et al. 2007; Van Engelsdorp et al. 2009; Bromenshenk 2010). The mechanism by which these two pathogens interact to potentially increase bee mortality is unknown. It may be that damage to gut epithelial and other host cells by the *N. ceranae* polar tube allows more robust entry of the virus. Alternatively, replication of *N. ceranae* in honeybee cells may cause a decrease in the bees' ability to ward off viral infections that normally could be controlled. These types of studies await isolation of the definitive virus.

Other than the presence of the IIV, only the concurrent absence of DWV, another RNA virus, was significant with respect to CCD. We recognize that the iflaviruses VDV-1 and Kakugo virus appear to be variants of DWV, whereas the dicistroviruses, Kashmir bee virus, acute bee paralysis virus, and IAPV are also closely related to one another (Baker and Schroeder 2008). Even closely related viruses may express different etiologies, and as such, we treated them as separate variables in our statistical analyses.

Virtually all of the bees from CCD colonies contained *Nosema* species and IIV, whereas IIV was not found in bees from packages imported from Australia nor in bees from colonies of the non-migratory, commercial bee operation in Montana. CCD has not been reported by the Australians (Anderson and East 2008) or the Montana beekeeper.

Since unapparent nonlethal infections by IIVs are common (Williams 1998), detection of IIV in some strong colonies and in the remnant young bee populations of collapsed colonies is to be expected. Presence of IIV in bees located in strong colonies may indicate a nonlethal infection, an early stage of the disease, or possible resistance to the virus in these colonies. In collapsed colonies, the presence or absence of IIV in remnant young bee populations is likely dependent upon the extent of infection in the colony and unknown pathophysiological factors that may affect persistence, colonization, and replication of the virus. Development of a PCR- or antibody-based assay to detect IIV in honeybees will allow us to rapidly detect and track the presence of IIV throughout the honeybee life cycle in a given colony.

Large amounts of IIV in failing colonies is consistent with an infection that proliferates in bees, but not necessarily to a degree that results in the iridescent coloration of infected bee tissues that is characteristic of IIV disease. The sustained high levels of IIV and *Nosema* peptides that occurred in the observation colony as the frequency of forager flights declined strongly implicates an IIV-6-like virus and *Nosema* as cofactors in CCD, since forager bee activity declined as pathogen loads peaked. In Spain, researchers have published studies linking *N. ceranae* to colony collapses in that country, yet we noticed that pest and disease surveys have reported the presence of an iridescent virus in colonies surveyed for mites and diseases (Anonymous 2000).

Because of their virulence, IIVs have been investigated as candidates for use as biopesticides (Bilimoria 2001; Marina et al. 2005; Becnel and White 2007). IIVs are large, icosahedra, double-stranded DNA viruses (Williams 1998, 2008). Of the many isolates reported from insects, only two, IIV-3 and IIV-6 (Williams 1998; Jakob et al. 2001; Delhon et al. 2006), have been subjected to complete genome sequencing and an additional 24 have been partially characterized (Tinsley and Kelly 1970; Williams and Cory 1994; Williams 1998). Historically, IIVs were numbered according to date of isolation (Tinsley and Kelly 1970; Williams and Cory 1994). Uniformly packed particle arrays (Webby and Kalmakoff 1998;

Williams 2008) of these viruses produce opalescent colors in the tissues of heavily infected hosts, particularly in insects in damp or aquatic habitats. These viruses have been shown to alter insect growth, longevity, and reproduction, and induce cell apoptosis (Cole and Morris 1980; Webby and Kalmakoff 1998; Williams 1998, 2008; Marina et al. 1999, 2003; Tonka and Weiser 2000; Lopez et al. 2002; Paul et al. 2007; Chitnis et al. 2008). In silkworms, IIV-1 can induce epidermal tumors (Hukuhara 1964).

Patent IIV infections are almost invariably lethal, but inapparent or covert infections may be common (Williams 1998, 2008). Inapparent infections may not be lethal, but may affect the reproduction and longevity of covertly infected hosts (Marina et al. 1999). IIV-3 is thought to be restricted to a single species of mosquito (Delhon et al. 2006; Williams 2008), although we found peptides close to those of IIV-3 in bees from an observation hive. These bee samples were handpicked with forceps, so we are confident that our observation hive samples did not include mosquitoes or other insects.

Other IIVs, such as IIV-6, naturally infect various species of Lepidoptera and commercial colonies of Orthoptera. There is evidence that hymenopteran endoparasitoids can become infected if they develop in an infected caterpillar (Lopez et al. 2002). IIVs have also been studied for control of mosquitoes (Marina et al. 2005; Becnel and White 2007) and boll weevil (Bilimoria 2001); the latter work examined the virus itself, with an emphasis on the proteins that it produces as the basis for a possible biopesticide. A U.S. Patent has already been awarded (Bilimoria 2001).

There is one known iridescent virus in bees. IIV-24, originally isolated from the Asiatic honeybee *Apis cerana*, severely affects bee colonies, causing inactivity, crawling, and clustering disease (Bailey et al. 1976; Bailey and Ball 1978; Verma and Phogat 1982). Proteomics could not identify IIV-24 in any of our samples because there are no IIV-24 sequences in the current databases. Thus, the identity of the IIV in our samples remains undetermined.

Based on the sequence data generated from MSP, the IIV identified appears to be closely related to IIV-6, possibly because this is the only IIV in the *Iridovirus* genus that has been completely sequenced. The major capsid protein represents ~40% of the total particle polypeptide and is highly conserved, so sequencing peptide fragments may frequently identify IIV-6 as being the most likely candidate (Tinsley and Kelly 1970; Webby and Kalmakoff 1998). This argument is reinforced by some results coming back as IIV-3, which is presently assigned to a different genus in the family (Chloriridovirus) and only reported to occur in a mosquito (Delhon et al. 2006).

There is little information about IIVs in bees, although there are historical reports associating IIVs with severe bee losses in India (Bailey et al. 1976; Bailey and Ball 1978; Verma and Phogat 1982), the United States (Camazine and Liu 1998), and possibly Spain (Anonymous 2000). In the 1970s, in northern India, almost every bee was infected with IIV-24, with 25%–40% annual colony loss (Bailey and Ball 1978; Verma and Phogat 1982). The disease was manifested by inactivity, clustering, and crawling sickness.

Transmission of IIV-24 is suspected to occur via eggs, feces, or glandular secretions in food (Bailey and Ball 1978; Verma and Phogat 1982). Evidence that IIV-24 was the cause of Indian bee losses was based on turquoise and blue iridescence seen in affected bees and tissues, serological tests, and microscopic examination of sick bees. IIV was the only recognizable parasite in all samples. IIV-24 was strongly correlated with coinfective *Nosema* species and tracheal mites in diseased colonies of *A. cerana* (Bailey and Ball 1978; Verma and Phogat 1982). Tracheal mites were found in some, but not all of the sick colonies (Bailey et al. 1976; Bailey and Ball 1978). The fat body was always attacked by the virus, and other tissues and organs, including the ovaries, were frequently infected (Bailey et al. 1976; Bailey and Ball 1978).

In addition, an iridescent virus has also been associated with mites, which may act as vectors, and has been implicated in bee losses in the United States. While investigating unusually high losses of bees in the northeastern United States, Camazine and Liu (1998) extracted a putative iridovirus from Varroa mites collected from a colony that perished 4 weeks later. They concluded that viral transmission within the colony might kill both mites and bees, but they were not able to discover the virus in time to determine whether bees in the colony were infected, and they were unable to purify the virus or determine whether the virus could be transmitted to bees by inoculation.

One or more species of external mites were suspected of being carriers of the virus in Indian bees (Bailey et al. 1976; Bailey and Ball 1978; Verma and Phogat 1982), as was also the case in the United States, with Varroa acting as the vector (Camazine and Liu 1998). The need for a better knowledge of the ecology of IIVs has been emphasized in order that preventive measures could be taken not only to offset damage to *A. cerana* but also to reduce the chance that *Apis mellifera* could become infected by this pathogen (Bailey and Ball 1978). Indeed, IIV-24 was experimentally inoculated and found to lethally infect *A. mellifera*, forming cytoplasmic quasi-crystalline aggregates of virus particles in cells of the fat body, hypopharyngeal glands, the gut wall, and proximal ends of the Malpighian tubules (Bailey et al. 1976; Bailey and Ball 1978).

These historical findings of IIV, mites, and *Nosema* species are intriguing since researchers studying both *N. ceranae* and CCD in Spain observed IIV-like particles in bee samples by electron microscopy (Anonymous 2000). The U.S. investigators studying CCD observed structures in thoraxes of bees described as "peculiar white nodules," resembling tumors that contained crystalline arrays (CCD Working Group 2006), similar to those described for IIV infections. Also, it appears that the IIV-6 genome encodes for one or more polypeptides that can produce insect mortality by inducing apoptosis without the need for viral replication (Paul et al. 2007).

The suspected source of *N. ceranae* in *A. mellifera* is the Asian bee *A. cerana* (Fries et al. 1996). This bee species is also known to be infected by Thai SBV and Kashmir bee virus. Kashmir virus was first detected as a contaminant in a sample of iridescent virus from India, as well as IIV-24 (De Miranda et al. 2004). The same virus was linked to bee losses in Canada in the early 1990s (Bruce et al. 1995). This suggests that perhaps not only the microsporidium *N. ceranae* but also other pathogens may have jumped from *A. cerana* to *A. mellifera*, as predicted by Bailey and Ball in 1978 (Bailey and Ball 1978).

It also implies that if Kashmir bee virus has been in North America for more than 20 years, so might IIV and *N. ceranae*. That would fit the time line of the first observations of this complex of pathogens, and of severe bee losses in India in the 1970s.

It also leads us to ask whether the first widespread losses of bees in the United States, described as Disappearing Disease in the 1970s (Wilson and Menapace 1979), may have been early outbreaks of CCD.

Our own work, described here, provides multiple lines of correlative evidence from MSP analysis that associate IIVs and *Nosema* with CCD in the United States. We conclude with results of laboratory inoculations of caged bees with IIV and *Nosema* that demonstrate the potential for increased lethality of mixed infections of these two pathogens. Our study strongly suggests a correlation between an iridescent virus, *Nosema* and CCD. Our inoculation experiments confirmed greater lethality of an IIV/*Nosema* coinfection compared to infections involving each pathogen alone. Future research using the specific strains of IIV isolated from infected bees will surely confirm whether a synergistic or additive interaction between these two pathogens results in the signs and symptoms of CCD.

The fact that IIV-6-inoculated bees experienced increased mortality in the presence of *Nosema* clearly strengthens the significance of all lines of evidence pointing to an interaction between an IIV and *N. ceranae*. Lack of a stronger effect by preparations containing IIV-6 may be due to the possibility that the IIV detected by proteomics is either a strain of IIV-24 or a strain of IIV-6 that is more specifically adapted to honeybees, and consequently more virulent than the strain of lepidopteran origin used in our inoculation experiments. It is, of course, critical to isolate the IIV from CCD populations, compare it to known IIVs and particularly IIV-24, and then challenge CCD populations with this strain. This work is in progress.

Moreover, we used a fairly low dose of IIV-6 and *N. ceranae* spores. For example, IIVs are generally not highly infectious by ingestion. Similarly, virulence studies on *N. ceranae* have reported using over 200,000 spores per bee in cage trials whereas we used a fourfold lower dose. It will also be interesting to test whether the interaction between IIV and *N. ceranae* is specific, or a general "stress" phenomenon that could also be reproduced by addition of *N. ceranae* and any additional bee virus.

In our studies, we applied six independent scenarios to the assessment of potential causes or markers of CCD and got the same answer, giving us confidence in the results, since this inference approach is approximately analogous to applying the same technique to six different assessments (Suter 2007). Our results also provide credibility to older, often overlooked work by others that associated IIV with bees, tracheal and Varroa mites, *Nosema* species, and severe bee losses. In our samples, Varroa mites were seen in many CCD colonies, but not in all. Importantly, our limited results do not completely fulfill the requirements of risk characterization, nor do they clearly define whether the occurrence of IIV and *N. ceranae* in CCD colonies is a marker, a cause, or a consequence of CCD. Our findings do make a strong case for a link between an iridescent virus and *Nosema* with CCD and provide a clear direction for additional research to answer these questions.

We anticipate that there may also be questions as to why IIV was detected in our study, but has not been found in any current published research on CCD. And, if these viruses were present, why were not they seen in infected tissues of the European honeybee, *A. mellifera*?

First, iridescent viruses have been seen before in *A. mellifera*, both in Europe and in the United States. Researchers in Spain reported seeing iridescent virus in honeybees (Anonymous 2000), and Camazine (Camazine and Liu 1998) saw a putative iridescent virus in Varroa mites following a collapse of colonies in the northeastern part of the United States in the 1990s. Also, inapparent infections by iridescent viruses may involve a low density of IIV particles in infected host cells (Tonka and Weiser 2000); so without sensitive techniques such as MSP, it is not surprising that infections in CCD bee colonies were previously missed.

The large number of IIV proteins that we identified, 139 in all, represents a significant fraction of the total IIV proteome. The recently published genome for IIV-6 (D'Costa et al. 2004) suggests a total proteome of 137 unique proteins. The 139 polypeptides identified for the IIV strain in our study must therefore represent a near complete sample of the total viral proteome belying any criticism that our identification of IIV may be a spurious consequence of accidental matching of a few peptide fragments.

We conclude that the IIV/*Nosema* association may be critical in honeybee mortality linked to CCD. Although viral diseases are currently manageable only by culling, *Nosema* infections are treatable with several current management techniques. We suggest that for beekeepers suffering from colony losses, disruption of the potential IIV/*Nosema* relationship using treatments that are available to control *Nosema* species may be one option to help reduce honeybee mortality. Again, whether this identified bee IIV and its potential interaction with *Nosema* species is the cause or marker of CCD is unknown, but our results clearly suggest that further research in this area is urgently required.

7.4.5.1 Wide Area Bee Sampling

We collected sample sets of adult worker honeybees from several areas and years: (1–2) two initial sample sets of adult honeybees from CCD colonies were obtained in 2006–2007 from 12 beekeeping operations from western, northeastern, and southeastern regions of the United States, (3) samples from packages of imported Australian bees provided a non-CCD 2007 reference, (4) bees sampled in 2008 from a large, non-migratory beekeeping operation in northwestern Montana with no history of CCD provided a second reference set, and (5) bee samples obtained in 2009 from a Florida apiary when 500 colonies suddenly collapsed constituted an independent CCD sample set by location and year. In each apiary investigated and sampled for this study, based on visible signs of CCD as described by the CCD Working Group (2006), samples of 200–500 bees were collected from each of nine colonies judged to represent the three most populous, three failing, and three collapsed colonies. Our team was part of the CCD Working Group that investigated and sampled the first reported colonies with CCD in the United States (CCD Working Group 2006). We later published an expanded description of the signs of CCD and variations that occur with season and geographical area, and we have continued to inspect and sample colonies showing signs of CCD from many areas of the United States from December 2006 to the present. As such, we are well familiar with the signs and stages of CCD.

Typically, the largest colony populations had 10–14 frames of adult bees or more, and two or more frames of brood. The collapsed colonies had less than a frame of bees, often no more than the queen and a fist-sized cluster of very young bees. Failing colonies were defined as those that had no more than half the number of bees as the most populous colonies. These colonies often had an excess of bees, and had far more bees just days or a few weeks before the samples were taken, according to the beekeeper accounts.

Additional reference bees were obtained from packages shipped from Australia to the United States and from the most populous to the weakest colonies from apiaries of a large, commercial beekeeping operation in Montana that is geographically isolated and has no history of CCD. In this case, the weakest bee populations were only about 20% smaller than the largest bee populations.

All of the CCD operations were large, migratory beekeeping businesses that transported bees across state borders and rented colonies for pollination of almonds in California. The migratory colonies sampled in 2006 and 2007 represented two different migratory routes, one from the east coast to California and the other from North Dakota to California. In addition, when in California, the east coast and the mid-western colonies were separated by ~400 km, so that there was no overlap of either the apiary locations or highways of these two different migratory routes. Bees were shaken directly into new, clean 1 quart ZiplocH or 1 L Whirl-Pac H bags. The bags were sealed, placed in a cooler with frozen gel packs, and shipped by overnight express to the U.S. Army ECBC laboratory. Bees were often alive when received and were analyzed immediately. In a few cases, bee samples were frozen and stored in a 280 μC freezer until analyzed.

Following the same sampling methods, we sampled a repeat of CCD in Florida, where 500 honeybee colonies started from packages in October 2008, collapsed in January 2009. As mentioned earlier, the beekeeper who owned the colonies had experienced CCD in 2006–2007, and had been in one of the original beekeeping operations sampled by members of the CCD Working Group. As in 2006–2007, the colonies suddenly collapsed, demonstrating the characteristic signs of CCD (CCD Working Group 2006; Debnam et al. 2008).

7.4.5.2 Time Sequence Bee Sampling

We also observed the progression of CCD in a collapsing colony in an observation hive, taking 18 bee samples of approximately 10–60 bees per sample interval, over a 3 month period, ending when only a queen and four workers remained.

In the spring of 2008, we lost more than 50 of our research colonies to CCD. We took the frames, queen, and the small, surviving population of young bees from one of these collapsed colonies, put them in a five-frame observation hive, and fed them sucrose syrup. This colony soon produced a second queen, and both queens coexisted in the same colony, one on each side of the glass hive, together producing a rapidly increasing combined population of bees. The forager bees had access to both syrup and to abundant food resources from the University of Montana campus, UM's arboretum, and surrounding residential flower gardens. By mid-summer, this bee colony collapsed for the second time. We then began to sample bees from those remaining in the hive.

The number of bees sampled at each time point varied with the health of the colony. We attempted to collect at least 60 bees per sample interval, until the end, when too few bees remained to take even 10 bees, which is the minimum sample size used for proteomics analysis. We also recorded forager flight activity and forager losses using a bidirectional digital bee counter mounted at the entrance of the observation colony.

7.4.5.3 Inoculation Experiments

We are working on isolating the IIV that infects CCD bees for use in inoculations to perform Koch's Postulates. For our preliminary experiments, and because of the high degree of similarity between the CCD-related IIV and IIV-6, based on the MSP data, we elected to conduct inoculation trials using IIV-6 and *N. ceranae* to observe how these two pathogens may interact.

Bees were obtained from non-CCD colonies with no detectable levels of *Nosema*, as confirmed by PCR, from the MSU apiary. *N. ceranae* was obtained from local colonies known to be infected by the microsporidian. The New Zealand strain of IIV-6 was obtained from Dr. James Kalmakoff, reared in *Galleria mellonella* larvae, and purified on sucrose density gradients as described previously (Henderson et al. 2001).

Following emergence from brood frames in an incubator, 1–3 day old bees were placed into sterile cardboard cups in a plant growth chamber with controlled temperature (28 µC, relative humidity, and light). Using a 10 mL pipette, each bee was inoculated by feeding it a total of 2 mL in sugar water containing one of four treatments. Only bees that ingested the entire inoculum were used.

The following treatments were given: (1) controls—sugar water/PBS 1:1, (2) *N. ceranae*—2 mL containing 50,000 spores, (3) virus—2 mL of 0.25 mg/mL IIV-6 suspension in PBS/sugar water 1:1 (0.25 µg virus), and (4) *N. ceranae* + virus—2 mL containing 50,000 spores + 0.25 mg of virus in PBS/sugar water 1:1.

Thirty bees were inoculated in each group and the experiment was repeated four separate times for a total of 120 bees in each group. Bees that perished 24 h after the inoculation were not included in the statistical analysis. Bees were then monitored daily for a period of 14 days. Dead bees were removed immediately upon discovery and frozen at 280 µC.

Dead bees from the inoculation experiments were analyzed by PCR and proteomics to detect and confirm infections by *N. ceranae* and virus. We used Kaplan–Meier curve analyses and log-rank test statistics to determine the significance of the mortality results.

7.4.5.4 MSP Protocols for Double-Blind Samples

Bee samples were homogenized in 100 mM of ammonium acetate buffer using a tissue homogenizer. The supernatant was filtered to remove large particulates, followed by ultrafiltration at 300 kDa. All filtered bee samples were lysed using an ultrasonication probe at settings of 20 s pulse on, 5 s pulse off, and 25% amplitude for 5 min duration. To verify cells were appropriately disrupted, a small portion of lysates was subjected to 1D gel analysis. The lysates were centrifuged at $141,006g$ for 30 min to remove all cellular debris. Supernatant was then added to a Microcon YM-3 filter unit (Millipore, USA) and centrifuged at $141,006g$ for 30 min. Effluent was discarded and the filtrate was denatured by adding 8 M urea and 3 mg/mL DTT and incubated for 2 h in an orbital shaker set to 50 µC and 60 rpm. A 10 mL volume of 100% ACN was added to tubes and allowed to sit at room temperature for 5 min. Tubes were washed using 100 mM ABC solution and then spun down at 141,006*g* for 30–40 min. The isolated proteins were then digested with 5 mL trypsin at a solution of 1 mg/mL (Promega, USA) in 240 mL of ABC solution +5 mL ACN. Digestion was performed overnight at 37 μC in an orbital shaker set to 60 rpm. Sixty microliters of 5% ACN/0.5% FA was added to each filter and vortex mixed gently for 10 min. Tubes were centrifuged for 20–30 min at 141,006*g*. An additional 60 mL 5% ACN/0.5% FA mixture was added to filter and spun. Effluent was then analyzed using the LC-MS/MS technique.

A protein database was constructed in a FASTA format using the annotated bacterial and viral proteome sequences derived from all fully sequenced chromosomes of bacteria and viruses, including their sequenced plasmids (as of September 2008) (Aebersold and Mann 2003; Dworzanski et al. 2006). A PERL program (http://www.activestate.com/Products/ActivePerl) was written to automatically download these sequences from the National Institute of Health NCBI site (http://www.ncbi.nlm.nih.gov).

Each database protein sequence was supplemented with information about a source organism and a genomic position of the respective ORF embedded into a header line. The database of bacterial proteomes was constructed by translating putative protein-coding genes and consists of tens of millions of amino acid sequences of potential tryptic peptides obtained by the in silico digestion of all proteins (assuming up to two missed cleavages).

The experimental MS/MS spectral data of bacterial peptides were searched using the SEQUEST (Thermo Fisher Scientific, USA) algorithm against a constructed proteome database of microorganisms. SEQUEST thresholds for searching the product ion mass spectra of peptides were Xcorr, ΔCn, Sp, RSp, and ΔMpep. These parameters provided a uniform matching score of all candidate peptides (Keller et al. 2002; Aebersold and Mann 2003). The generated outfiles of these candidate peptides were then validated using the peptide prophet algorithm.

This validating and verification approach uses an expectation-maximization algorithm as described by Keller et al. (2002), the creators of PeptideProphet. The algorithm calculates a statistical score that reflects the confidence of the match to each peptide identified. Peptides identified are eliminated if they are below a selected threshold. In our case, the threshold was set at 95%. Peptides that were identified with less than 95% confidence were removed from the final data set.

Peptide sequences with probability score of 95% and higher were retained in the data set and used to generate a binary matrix of sequence-to-microbe assignments. The binary matrix assignment was populated by matching the peptides with corresponding proteins in the database and assigned a score of 0 (no match) or 1 (match). The column in the binary matrix represented the proteome of a given microbe and each row represented a tryptic peptide sequence from the LC-MS/MS analysis.

Bee samples were identified with the viruses/bacterium/fungi proteome based on the number of unique peptides that remained after removal of degenerate peptides from the binary matrix. This approach was successfully used for the double-blind characterization of non-genome-sequenced bacteria by MSP (Jabbour et al. 2010a,b).

Proteomics identified peptides described from a variety of bee viruses, as well nine species of *Nosema*: *N. apis*, *N. bombycis*, *N. locustae* (now known as *Antonospora locustae*), *N. tricoplusiae*, *N.* BZ-2006B, *N.* BZ-2006d, *N. granulosis*, *N. empoascae*, *N. putellae*, plus a collection of unspecified *Nosema* peptides. At the time that our proteomics analyses were conducted, the *N. ceranae* genome sequence was not available and only one *N. ceranae* sequence was available in the database.

It is almost certain that the diversity of *Nosema* represented in the proteomics results was not attributable to multiple infections by all the species identified. Rather, the taxonomic diversity in the data reflects historical precedent in the *Nosema* research that added different proteins to the genomic and proteomic libraries. Assuming that the *Nosema* proteome described in our data indicated one or at most a few species, we entered total peptide counts for each species into a hierarchical cluster analysis using average chi-squared distance between species. The analysis produced two primary groupings of *Nosema* peptides: Group 1 which contained *N. apis*, *N. bombycis*, and *N. locustae* and Group 2 which contained all of the remaining *Nosema* species.

We performed forward, stepwise discriminant analysis on square-root or log-transformed pathogen counts for *Nosema* species and for all of the bee virus species. Counts were calculated by weighting each pathogen occurrence by the total number of its peptides that were detected. The specific transform performed on each variable was the one that best normalized the distribution of individual variables.

Use of peptide counts as a weighting factor stems from the observation that as total pathogen titer in a sample increases, the number of different peptides that can be identified by proteomics increases in a predictable manner (Keller et al. 2002). Thus, the number of peptides observed for each pathogen served as a relative measure of its abundance in the sample.

Four colony groups were discriminated: strong, failing, collapsed, and the Montana reference group. The selection method for variable entry was largest univariate *F*-value; *F* to enter was set at 2.0. Equal probability of group membership was assumed.

The analysis was completed after two steps including only IIV-6 and DFW as significant discriminating variables (final Wilks' lambda = 0.679; $F = 2.881$; $df = 2, 54$; $p = 0.031$; Table 7.2). None of the *Nosema* groups were selected for the discriminant functions, but *Nosema* 1 was strongly correlated with IIV; the pooled within groups correlation matrix from which the discriminant functions were extracted showed that the highest among groups correlation was between IIV and *Nosema* 1 ($r = 0.89$). Because IIV and *Nosema* 1 conveyed the same discriminating information, only one was included.

7.4.6 COMMENTS

It should be noted that the database used in this study included the honeybee proteome that was available at the time, and also the published proteomes for plants, vertebrates, and other invertebrates as a guard against misidentification. The computational method employed has been verified and did not inflate FDR; to the contrary, it kept FDR below 5%. Also, it has been clearly demonstrated by blind analysis of specially prepared samples of mixed known pathogens that ABOid™ correctly identifies microbes without error.

This said, there were several hundred other microbes that were also detected from the honeybees. Some of these microbes were found to be those used by researchers at local sites within flight range of the various honeybee colonies. These were verified by taking with the various scientists. It was interesting that complete lists of these microbes could be complied complete to strain. This remains an interesting area of investigation and further study.

The methods used with the honeybees would also be useful for monitoring the microbe health of an area. It could be expected that there are useful microbes present and a change in this list may be interesting as it relates to changes in the environment.

8 Survey of Commercially Available MS-Based Platforms Suitable for Bacterial Detection and Identification

Michael F. Stanford

CONTENTS

8.1   INTRODUCTION

This is a moving mark, as new platforms arrive on the scene at regular intervals. This survey, therefore, is a beginning and serves to demonstrate many different platforms that may be of service to the detection and identification of microbes. The question as presented several times in this book is that it depends on the application. Simple detection, without the need for a thorough identification to strain, may be utilized on one or more of the more simple platforms. Simple identification to genera or even types of bacteria, some fungi, and a few virus types may be all that is required for many applications. Complex and detailed analysis for identification of strain or unknown microbes may indicate the most sensitive of the platforms.

One clear point is that as improvements in all these platforms occur and as they become smaller, faster, and less expensive, they all have greater potential. Combine this with smaller and faster computers which we see on a regular basis. The design, simplification, and miniaturization of sample-processing steps allow for the visualization of a portable device. Since the software is portable to many platforms, the potential is vast.

The subject of software is important. First, because it is frequently the defining means that is used by each of these platforms for analyzing a file. That is the information from the MS platform is analyzed to report some sort of result. Software used to determine pattern recognition is going to be limited to those applications. Likewise, software limited to other means of analysis will limit that platform to applications that the software is capable of processing. Do not expect detection and identification of microbes if the software being used is for chemicals. Fortunately, the software is frequently portable between types of instruments. The MS platform may be capable of resolving the mass data needed for microbe identification; it just needed the correct software.

Table 8.1 is a list of those platforms that are commercially available at this writing. A short review of their attributes may give an indication of their utility.

8.2   CBMS BLOCK II

The CBMS Block II is a new and improved system for the detection and identification of chemical and biological warfare agents (CBWAs, Figure 8.1). This system is currently being developed for the U.S. Army's Soldier and Biological Chemical Command by the Oak Ridge National Laboratory and their industrial partner,

TABLE 8.1
Commercial MS Platforms

	Manufacturer/Developer	System/MS Technology	Method/Specificity	Automation/Processing Time
1	Oak Ridge National Laboratory	CBMS-Qit	Profiles of fatty acids/low specificity	Bioaerosol/minute
2	Dephy Technologies, Inc. (Montreal, Canada)	Py-MS	Analysis of ionized cellular pyrolysis products/low specificity (fingerprints)	Automated/minute
3	Syagen Technology, Inc. (Tustin, CA)	Py-GC-Qit-TOF MS	Analysis of ionized cellular pyrolysis products/low specificity (fingerprints)	Automated/minute
4	TNO Defense, Security and Safety (Rijswijk, The Netherlands)	MALDI-ATOF-MS	Spectral analysis of ionized single bioparticle components	Automated/second
5	Lawrence Livermore National Laboratory (Livermore, CA)	BAMS	Low MW mass signatures of microbial particles/low specificity	Automated/minute
6	Middle Atlantic Mass Spectrometry Laboratory–Johns Hopkins University (R. Cotter)	Miniaturized TOF-MS-MALDI	Spectral analysis of ionized cellular components/microbial fingerprinting	Manual/minute
7	Science and Engineering Services, Inc. (SESI, Columbia, MD)	Wide-spectrum Bio-ID-AP-MALDI	Sequencing of tryptic peptides from microbial proteins/high specificity	Automated/minute
8	Johns Hopkins University-Applied Physics Laboratory (Laurel, MD)	Suitcase TOF-MS analyzer/MALDI	Spectral analysis of ionized cellular components/microbial fingerprinting	Manual/minute
9	Johns Hopkins University-Applied Physics Laboratory (Laurel, MD)	Bio-TOF-MS	Spectral analysis of ionized cellular components/microbial fingerprinting	Full automation/minute
10	Waters Corp. (Milford, MA)	V-MALDI-TOF-MS/ Microbe-Lynx software	Spectral analysis of ionized cellular components/microbial fingerprinting	Manual/3 minute
11	IBIS Biosciences, Inc./Bruker Daltonics, Inc.	TIGER/V-MALDI	TIGER	Fully automated/hour
12	Bruker Daltonics, Inc. (Billerica, MA)	BioProfiler/V-MALDI	Spectral analysis of ionized cellular components (microbial fingerprinting)	Fully automated/minute
13	Purdue University, West Lafayette (R.G. Cooks)	Rectilinear ion trap (ESI-DESI)	Spectral analysis of ionized cellular components	To be determined

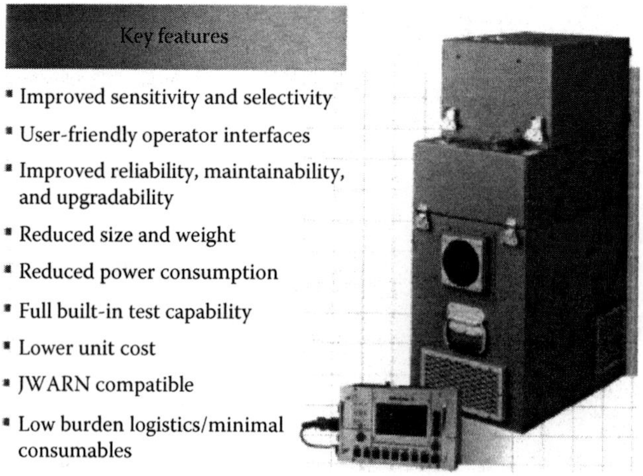

Key features

- Improved sensitivity and selectivity
- User-friendly operator interfaces
- Improved reliability, maintainability, and upgradability
- Reduced size and weight
- Reduced power consumption
- Full built-in test capability
- Lower unit cost
- JWARN compatible
- Low burden logistics/minimal consumables

FIGURE 8.1   CBMS (Block II).

Orbital Sciences Corporation. The CBMS Block II consists of a MS module, sample introduction module, biosampler module, and a soldier display unit. The biodetection system is based on direct sampling and thermolysis/derivatization of biological particulates. This "dry" system minimizes the logistical burden and operating costs. This system offers significant reductions in weight, size, and power consumption over the current CBMS and other systems. The CBMS Block II is designed for use in reconnaissance vehicles and other mobile detection systems.

The CBMS II is based on ion trap MS techniques for the detection and identification of CBWAs. It is being developed to satisfy the agent detection and identification requirements for platforms such as the Stryker Nuclear Biological and Chemical Reconnaissance Vehicle (NBCRV) and the Joint Service Lightweight NBC Reconnaissance System (JSLNBCRS). The CBMS II is intended to meet current CB threats and, depending on the platform, detect and identify toxic industrial chemicals (TICs) and nontraditional agents (NTAs).

The CBMS II maintains the ion trap mass scan ejection strategy used in the CBMS I. This strategy employs nonlinear fields to rapidly eject ions from the trap and allows air to be used as a buffer gas, thereby eliminating the need for compressed gases. The CBMS II also continues the use of fatty acid analysis for BA detection.

The bioconcentrator used by the CBMS II employs a novel "opposed jet design" that improves sensitivity and reduces the air intake flow from 1000 to 330 L/min. The pyrolysis assembly uses TMAH for in situ derivatization which is held in a 25 mL reservoir, sufficient for five 72 h missions. The derivatization of the carboxylic end of the fatty acids reduces polarity and increases sample transmission of primary biomarkers. The detector can be operated in several modes:

- Electron ionization (EI)—Full-scan EI is used for CAs and unknown chemicals.
- Chemical ionization (CI)—Full-scan CI is used for CBA detection.

- Tandem MS (MS/MS)—Precursor/product ions used for highly selective identification for CBAs.
- Multi-scan functions (MSF)—Multiplexed combination of full-scan MS and MS/MS for automatic collection of MS data and agent identification.

8.2.1 BASELINE SENSITIVITY

Using the application of the full bio-algorithm, with liquid agent injected into the unit, the system correctly identified ≥90% for all but one agent in the classification test. Only one agent was tested in aerosol form (*Erwinia herbicola*) at 25 ACPLA; the system correctly identified this agent 80% of the time in the classification tests.

8.2.2 PERFORMANCE

- BW threat agents (bacteria, toxins, and viruses)—25 agent-containing particles/L of air (ACPLA) detection and identification time biological, <4 min
- Size ($H \times W \times D$) 5.8 ft^3 ($36'' \times 20'' \times 14''$)
- Total system weight (including biosampler) 170 lbs (77 kg)
- System power peak power, <1000 W; operating, <500 W
- Input voltage 20–31 VDC
- Temperature range
- Operating −25°F to 120°F
- Storage −60°F to 160°F
- Humidity (operating) 5%–95% RH
- Shock and vibration MIL-STD-810E
- Electromagnetic interference MIL-STD-461
- Mean time to repair (MTTR) < 30 min
- Data storage capable of storing and maintaining 72 h of continuous operational data in memory

8.3 TOF BACTERIA ANALYZER USING METASTABLE SOURCE IONIZATION OF PYROLYSIS PRODUCTS (PY-MAB-TOF-MS)

Dephy Technologies Inc., 3552, St-Patrick Street, Montréal, Québec H4E 1A2 (CA); Inventors: Michel J. Bertrandt and Olivier Peraldi.

The Dephy MAB-TOF-MS uses reflectron ion analyzer to produce unit mass resolution up to *m/z* 2000. The MS has a small footprint of $24'' \times 28''$ because it was designed as a field-portable instrument inside a small vehicle (Figure 8.2). It is possible to obtain very good quality pyrolysis Ar* MAB mass spectra from 50,000 cells.

The instrument collects 16,000 full range spectra per second, thus increasing its sensitivity several hundred times in comparison to scanning quadrupole instruments. A significant practical limitation of this reflectron TOF mass analyzer is the number of independent, interactive variables that must be adjusted to tune and calibrate. However, the development of a fully automated auto-tune program should solve this problem in the near future.

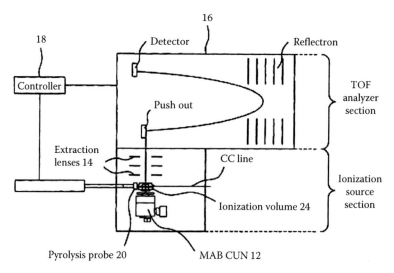

FIGURE 8.2 TOF bacteria analyzer using metastable source ionization of pyrolysis products. A microbial sample is prepared, placed in a pyrolyzer where it is pyrolyzed with a selected temperature program to provide the pyrolyzed product of a high-dalton mass range. The product is ionized using metastable atoms, which results in efficient ionization with little fragmentation. The metastable atoms are generated using a generator that provides a beam of metastable atoms which is basically free from ions. The ionized product is then analyzed using a high acquisition rate mass analyzer, such as a TOF analyzer.

The average spectrum is defined by integrating across the pyrogram peak from the time it first rises above baseline until it returns to within 10% of the baseline. This covers a period of around 45 s during a typical data acquisition. The Dephy data-processing package offers several proprietary noise filtering options that improve spectral reproducibility and thus help in the classification of samples.

8.4 PY-GC-QIT-TOF-MS SYSTEM FOR CLASSIFICATION OF BWAs

A Syagen Technology, Inc. (Tustin, CA), has adapted the advanced photoionization (PI)-Qit-TOF-MS technology (Figure 8.3) into a field-portable system called the FieldMate. Developed for high-performance, real-time analysis in the field, the FieldMate permits direct air and liquid sampling using the PI source. It also accommodates a flash GC interface for more detailed analyses using an optional EI source. The FieldMate (see Figure 8.4) was developed for the DoD to perform rapid and continuous monitoring for chemical/biological defense (CBD). The FieldMate equipped for biological weapons detection employs a Py-GC front end for breaking down BWAs, such as bacteria and viruses, into chemical signatures and analyzing them in two dimensions by GC-MS (Figure 8.5). The use of PI and Qit-TOF is crucial for obtaining clear and distinct signature information. Target specifications are gathered in Table 8.2.

Ion optics

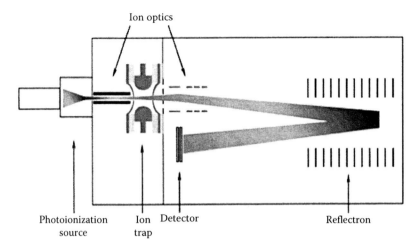

Photoionization source   Ion trap   Detector   Reflectron

FIGURE 8.3 Schematic representation of the PI source coupled with quadrupole ion trap to store ions. The stored ions are mass analyzed using mass-selective instability scanning and are pulsed into a reflectron TOF mass analyzer.

8.5 FIELDABLE BIODETECTION SYSTEM USING MALDI-AEROSOL TOF-MS OF SINGLE PARTICLES

Researchers at the Delft University of Technology and TNO Defence, Security and Safety (Rijswijk, The Netherlands) developed a technique that combines fluorescence pre-selection and MALDI-TOF-MS to obtain mass spectra from single biological aerosol particles (Marijnissen et al. 1988; Kievit et al. 1996; Stowers et al. 2000; Van Wuijckhuijse et al. 2005). Mass spectra from a single or a few individual aerosol particles can be obtained within seconds, enabling rapid identification of the aerosol material. The system incorporates an evaporation/condensation flow cell that allows in-line matrix coating of the aerosol particles (Figure 8.6). The coated particles are separated from the surrounding air using aerosol beam techniques. The passage through the focal point of a UV laser induces fluorescence in biological materials. This effect can be used for pre-selection, which enables the preferential analysis of aerosol particles of biological origin. Detection and sizing are achieved by collecting the light scattered when the particles pass through the focal points of a second laser beam. This event also triggers the desorption laser. The matrix of UV-absorbing material ensures efficient ion formation, and the resulting mass spectra can be used to identify the aerosol material.

The aerosol MS is designed to detect single particles in the size range of 0.5–20 μm in diameter. The concentration of these particles in the atmosphere is highly variable, but is commonly on the order of 100 particles per liter of air. If all particles in this size range are analyzed, this implies that, on average, 10,000 particles must be analyzed for every agent-containing particle. Given the maximum rate of operation of the system, 100 Hz, only a few agent-containing particles would be analyzed during any 5 min period. To address this situation, a mechanism has been developed whereby the system

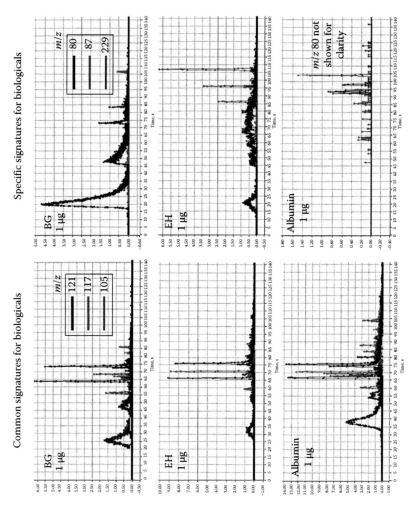

FIGURE 8.4 A fieldable Py-GC-Qit-TOF-MS for the classification of BWAs developed by the Syagen Technology Inc.

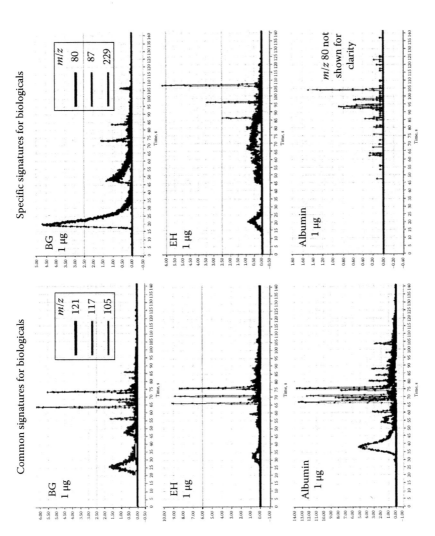

FIGURE 8.5   Common and specific signatures for biologicals obtained by Py-GC-PI-MS (Syagen Technology, Inc.).

TABLE 8.2
Target Specifications for Syagen's CB MS Screening System

Performance Characteristics		Operational Characteristics	
Sensitivity—particle	<10 ppb), 1 ACPLA for BWA	Start-up time	10 min
Specificity	>10^4 clutter suppression	Environment	−5°C to +60°C
Cycle time	2–3 min (BW)	Flexibility	CW and BW detection
Dynamic range	4-decade		

System Characteristics		Physical Characteristics	
Sampler	Particle concentrator	Size	14×16×18 in
			(w/o concentrator)
Collection flow	700 L/min	Weight	50 lbs (w/o concentrator)
Mass analyzer	Qit-TOF	Power	24 VDC, 6 A (115 VAC, 2 A)
Mass range	20–1000 amu		
Mass accuracy	$\Delta m < 0.1$ amu		

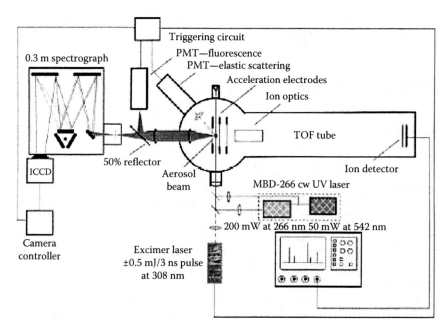

FIGURE 8.6 Schematic representation of the fieldable biodetection system using MALDI-ATOF-MS of single aerosol particles.

reacts only to particles that are very likely to contain bacteria, which are known to emit fluorescence when excited with UV light. The particle detection system has been designed to use this property to pre-select bacteria from a mixed aerosol. The system detects particles based on light scattering emitted as the particles pass two detection laser beams, with the time between scattering events related to particle size. One of these beams is composed of 266 nm light. As the particles pass this beam, emitted

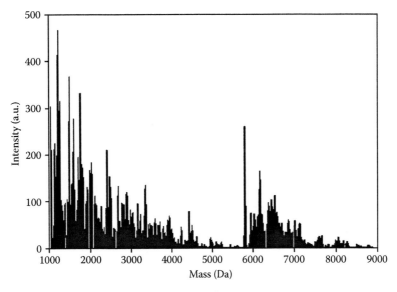

FIGURE 8.7   Aerosol mass spectrum of *B. subtilis* var *niger.*

fluorescence in the range of 300–450 nm distinguishes the particles existing of bacteria from the larger background population of particles. By limiting the subsequent mass spectral analysis to fluorescing particles, this system performs an efficient pre-selection of the bacteria present in the sample. An example of a mass spectrum obtained during sampling of aerosolized *B. subtilis* var. *niger* spores is shown in Figure 8.7.

8.5.1   Characteristics of MALDI-MS Performed on Aerosol Particles

1. Process
   - Aerosol sampling (up to 100× concentration of the aerosol particle density)
   - Sample preparation (online matrix coating)
   - Transport of aerosol to high vacuum (selection of particles with a "desired" size range)
   - Pre-selection of bioparticles (266 and 532 nm continuous wave beams, and pulsed 337 nm)
   - Ionization (adding matrix for absorption of UV laser pulse and charge transfer to analyte)
2. Detection
3. Integration. Laboratory setup migrated to a platform suitable for installation into (armored) vehicles.

8.6   BIOAEROSOL MS FOR DETECTION OF BIOLOGICAL PARTICLES IN AMBIENT AIR

Researchers from the Lawrence Livermore National Laboratory (Livermore, CA) have developed the Bioaerosol MS (BAMS) system for the real-time detection and identification of biological aerosols. Particles are drawn from the atmosphere directly into vacuum and tracked as they scatter light from several continuous wave lasers. After tracking, the fluorescence of individual particles is excited by a pulsed 266 nm or 355 nm laser. Molecules from those particles with appropriate fluorescence properties are subsequently desorbed and ionized using a pulsed 266 nm laser. Resulting ions are analyzed in a dual-polarity MS.

The creation and observation of higher mass ions are needed to enable a higher level of specificity across more species. A soft ionization technique, MALDI, is being investigated for this purpose.

8.6.1 DESCRIPTION OF THE SYSTEM

A schematic representation of the system is shown in Figure 8.8. In this system aerosol is first drawn into a virtual impactor, which concentrates the particles falling within the respirable size range (~1–10 μm). The concentrated aerosol is then sampled into the vacuum system of the BAMS instrument. A supersonic expansion into vacuum focuses the particles into a vertically orientated beam. The rest of the BAMS system is composed of stages that generally decrease in speed, but increase in specificity along the particle trajectory. The tracking stage determines a particle's aerodynamic diameter. After that, one or more stages examine a particle's intrinsic fluorescence. During the first deployment, 266 nm laser pulses were used for excitation. During the second, 355 nm pulses were used. In newer instruments, both stages can be installed in series along with other stages utilizing other nondestructive analysis techniques. Ultimately, molecules from the particles are desorbed and ionized with a single 266 nm laser pulse and a dual-polarity mass spectrum is collected.

The instrument utilizes a commercial, dual-polarity, reflectron TOF-MS from TSI Inc. The reflectron allows a longer flight path to be contained within a given sized flight tube and it helps correct for any spread of initial kinetic energies of

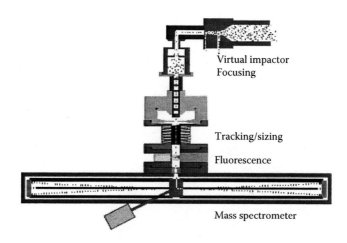

FIGURE 8.8 Schematic representation of the BAMS system.

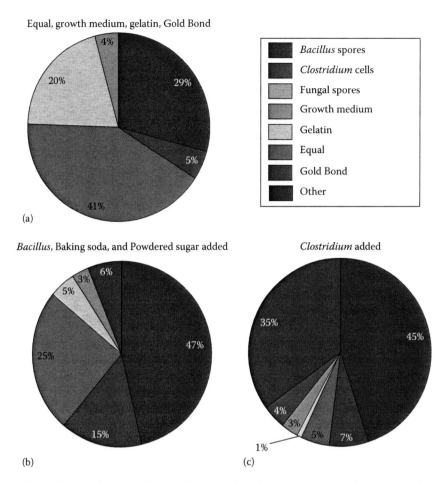

FIGURE 8.9 Pie charts showing real-time classifications of mixed aerosols based on BAMS results. It is worth noting that fungal spores were not present in any of the mixtures and so were not detected. (From Fergenson, D.P. et al., *Anal. Chem.*, 76(2), 373, 2004.)

ions. However, recently, the new linear MS (which does not have a reflectron) was designed to greatly improve the efficiency of ion transport and demonstrated record-breaking sensitivity.

The classification of mixed aerosols using the reflectron TOF-MS system is shown in Figure 8.9. For this purpose the BAMS real-time recognition algorithm was programmed to detect individual particles of seven analytes, and in real-time recognition tests, the system detected each with absolute specificity, illustrating the uniqueness of the spectral signatures.

The sensitivity varied by sample. *Bacillus* spores were recognized 92% of the time; Gold Bond powder, 91%; growth medium, 86%; Equal sweetener, 78%; fungal spores, 56%; and Knox Gelatin, 46%. Spectra that were unrecognized were classified as other. In this case, the laser power can be optimized to identify *Bacillus* spores (Fergenson et al. 2004).

8.7   MINIATURIZED TOF-MS FOR BIOAGENT DETECTION AND IDENTIFICATION

The Middle Atlantic Mass Spectrometry Laboratory has developed a miniaturized TOF-MS that will be used in environmental, bioagent detection, and diagnostics applications (Figure 8.10). It has a "linear" design; the ions are focused to a detector located within a line of sight of the ion source; and no reflectrons are used. It uses pulsed ion extraction and other time-dependent electric fields (Figure 8.11). These mass analyzers have flight paths of ~3 in. with mass resolutions of up to one part in 1200 and mass ranges over 66 kDa. The use of nonhomogeneous electric fields is also proposed to further improve that performance. The application of this instrument for the analysis of bacterial spores is shown in Figure 8.12.

8.7.1   INSTRUMENT DESIGN

The dimensions of the ion source and extraction regions are 0.13 and 0.22 in., respectively, while the mass analyzer consists of a 3 in. long stainless steel tube and an insulating spacer. Samples are deposited on a small sample plate mounted on an XY sample stage. The ion detector is a Hamamatsu microchannel plate with a 4 mm channel diameter. The laser is a 337 nm pulsed nitrogen laser, which is attenuated with a variable neutral density filter. A digital delay generator is triggered by the laser pulse and, after a preset delay, generates an ion extraction pulse which is amplified

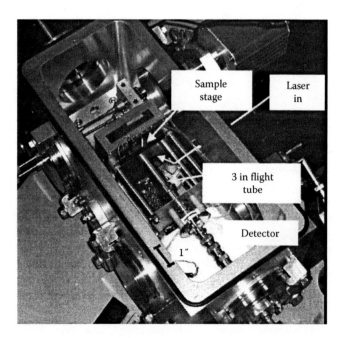

FIGURE 8.10   Photograph of the miniature MS with 3 in. analyzer.

Miniaturized linear, 3 in., pulsed TOF mass spectrometer

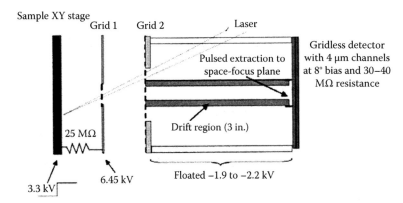

FIGURE 8.11 Schematic representation of the miniature instrument indicating the applied potentials.

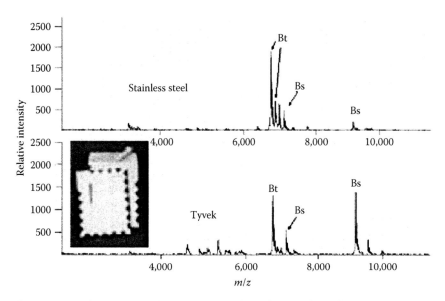

FIGURE 8.12 Comparison of mass spectra of a mixture of *B. thuringensis* (Bt) and *B. subtilis* (Bs) spores obtained from the stainless steel sample stage and adsorbed onto a Tyvek Envelope. A sinapinic acid solution was used as matrix and 10% trifluoroacetic acid for spore lysis.

by a fast high-voltage transistor switch. The instrument is enclosed in a $13'' \times 6'' \times 6''$ stainless steel chamber. Mass-correlated acceleration has been a more recent implementation on a miniature instrument and showed encouraging performance results for peptide analysis.

8.7.2 Performance for the Analysis of Peptides, Proteins, and Bacterial Spores—Mass Resolution

The delay time–optimized TOF mass spectrum for ACTH showed the resolution of the larger peak at m/z 4542, which had a FWHM of 2.0 ns and a resulting mass resolution of 1200. Because delay times are mass-dependent, for a peak at m/z 3660, a FWHM of 2.5 ns and a mass resolving power of 870 were observed. Overall, the 2.0 ns peak width for a molecular ion cluster spanning approximately 3–4 isotopic species at half height is close to the maximum resolution that can be expected for this instrument. Figure 8.12 shows the mass spectra of biomarkers that demonstrate the wide range of m/z that can be recorded on the 3 in. mass analyzer. These increased peak widths reflect in part the increase in the number of isotopic species that make up the molecular ion distribution. Moreover, it has the ability to record ions of high mass as was demonstrated in the case of mass spectra of the C fragment of tetanus toxin (52 kDa) and bovine serum albumin (66 kDa).

8.7.3 *Bacillus* Spore Identification via Proteolytic Peptide Mapping

A rapid screening method for the presence of *Bacillus* spores based on the analysis of peptides has been developed and tested with a miniaturized MALDI-TOF-MS (Figure 8.13). A limited set of tryptic peptides was generated in situ following selective solubilization of the small, acid-soluble protein (SASP) family from spore samples on the MALDI sample holder. To facilitate species identification, a compact database was created comprising masses of the tryptic cleavage products generated in silico from all *Bacillus* and *Clostridium* SASPs whose sequences are available in public databases. Experimental measurements were matched against the custom-made database, and a published statistical model was then used to evaluate the probability of false identification. The successful implementation of this approach is attributed to the fact that spectra of peptides, when compared to intact proteins, offer the advantages of more sensitivity, better reproducibility, and more accurate mass analysis. Moreover, further improvements are possible because the bioinformatics approach applied did not take advantage of specific peptides known to be unique to each species (English et al. 2003).

8.7.4 Rapid Detection and Identification of Microorganisms

Recently, SESI researchers reported development of a high-throughput AP-MALDI-MS-based assay that offers important benefits such as seamless integration of sample preparation with MS and fast overall analysis, on the order of minutes (Oktem et al. 2007) (Figure 8.14). For initial studies, they modified a Multiprobe II laboratory workstation (Perkin Elmer LAS, Shelton, CT) by including customized liquid

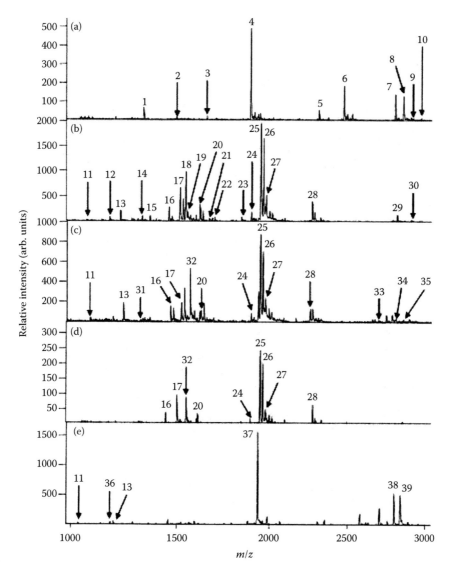

FIGURE 8.13 MADLI spectra of the tryptic digests generated in situ from (a) *B. subtilis* 168, (b) *B. anthracis* Sterne, (c) *B. cereus* T, (d) *B. thuringiensis* subsp. *kurstaki* HD-1, and (e) *B. globigii* spores and analyzed with the miniaturized TOF-MS. Peaks that were matched to peptides in the SASP database are numbered 1–39. Peaks that occur in more than one spectrum carry the same number. (From English, R.D. et al., *Anal. Chem.*, 75(24), 6886, 2003.)

handling by BioHIT (Helsinki, Finland). It was used for sample preparation on C-18-functionalized AP-MALDI target probe at 50°C–60°C. The customized system includes (a) automated plate transfer from the sample-processing workstation to the MS, (b) fast actuators to reduce sample preparation time, (c) compact cooling mechanism to increase duration of unattended operation, and (d) expansion possibility to

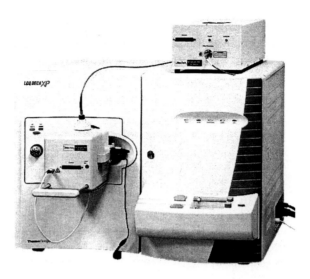

FIGURE 8.14   AP MALDI installed on the thermo Finnigan ion trap MS.

include different types of liquid handling and alternative sample introduction possibilities. The workstation was integrated with Varian 500-MS or Thermo LCQ
DecaXP quadrupole ion trap MS (Figure 8.15). The on-probe sample-processing
protocol is depicted in Figure 8.16.

An example of spectra and database search results obtained during the analysis
of bacterial cells is shown in Figure 8.17, and for a viral sample, in Figure 8.18. The
spectra were recorded with an LCQ-Deca XP ion trap MS (Thermo Finnigan, San
Jose, CA) equipped with an AP-MALDI source (MassTech, Columbia, MD) with a
337 nm nitrogen laser. The laser power was optimized between 70 and 120 μJ, with
a laser firing rate of 10 Hz. Laser shots were rastered manually across the sample.
Spectra were accumulated for 1 min intervals and averaged.

8.8   SUITCASE TOF-MS

Researchers from the JHU-APL (Columbia, MD) developed a Suitcase
TOF-MS for field applications (Figure 8.19). The Suitcase TOF was designed for
a first-responder type of application in which a small instrument must be easily transportable to a remote location for the testing of a suspect material. The
sample to be analyzed is co-deposited with a UV-absorbing substance (matrix)
onto a probe and inserted into the source of the analyzer. A "person in the loop"
is also envisioned for this instrument—a minimally trained individual who can
perform a simple sample workup with a disposable test kit, move the sample into
the instrument, and start the analysis sequence. The inventors claim that test kits
can be developed for solid, liquid, and gaseous samples so that the result can be
obtained within a few minutes.

The Suitcase TOF-MS (Figure 8.19) has four major subsystems—the vacuum
system, optical system, source/analyzer, and electronics/data system.

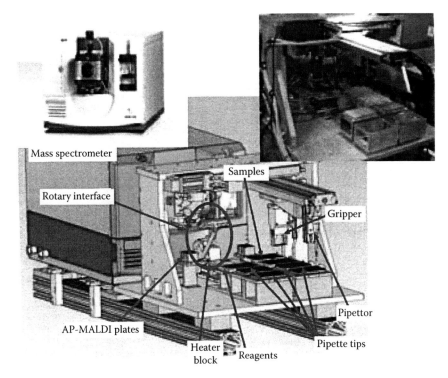

FIGURE 8.15 3D model and the actual picture of the automated system.

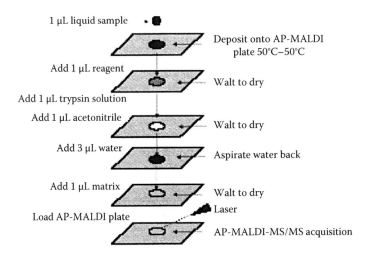

FIGURE 8.16 On-probe sample-processing protocol.

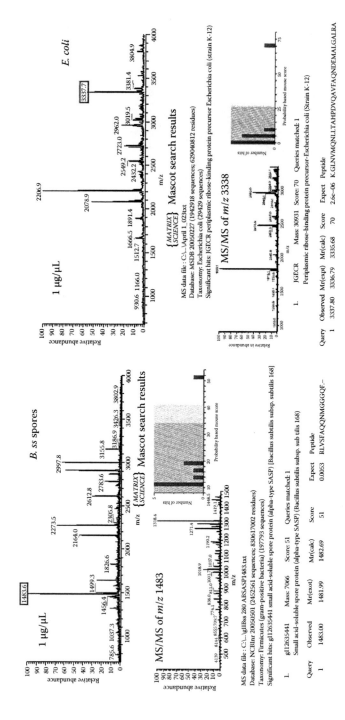

FIGURE 8.17 AP-MALDI analysis and identification of environmental microorganisms. The bacterial cells used were *B. subtilis* subsp. *subtilis* (*B. ss*) spores and *E. coli*. Bacterial proteins were released and digested with immobilized trypsin beads. Matrix: CHCA.

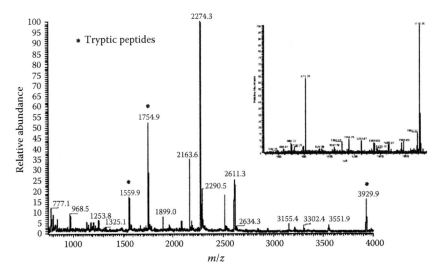

FIGURE 8.18   AP-MALDI-MS and MS/MS spectrum (inset) of MS2 bacteriophage capsid protein after on-probe sample processing. Concentration of MS2 was 2.5×10^6 pfu/sample.

8.8.1   VACUUM SYSTEM

The vacuum system consists of a 52 in.^{3} coffin-type chamber. A combination of turbomolecular/drag pump is used to evacuate the chamber into the low 10^{-6} torr range with a pumping speed of 10 L/s and a mass of 2.5 kg. This type of combination pump can exhaust into the relatively low vacuum of a diaphragm pump capable of operating at 1.5 torr and 4.8 L/min. The chamber pressure is measured with a wide-range vacuum gauge (combination Pirani/cold-cathode gauge).

8.8.2   OPTICAL SYSTEM

The optical system of a typical MALDI-TOF-MS is designed to deliver a series of short UV laser pulses to the source region of the TOF analyzer. The sample to be analyzed is co-deposited with a UV-absorbing substance (matrix) onto a probe and inserted into the source of the analyzer at the focal point of the optical system. Both analyte and matrix molecules are desorbed, ionized, and accelerated into the TOF analyzer. The APL Suitcase TOF uses fiber optics to deliver the energy from a pulsed nitrogen laser featuring a small, sealed plasma cartridge (140 µJ per pulse, 5 ns pulse width, 337 nm wavelength, and 1–20 Hz pulse rate). A small lens pigtailed to the end of the second fiber focuses the light through a vacuum window and onto the center of the sample probe. A simple sliding gimbal mount is used to align the beam and set the spot size at the sample. Finally, a fast photodiode mounted near the exit aperture of the laser detects scattered light to be used as the start trigger for the TOF measurement.

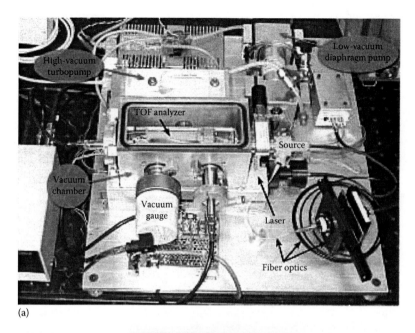

(a)

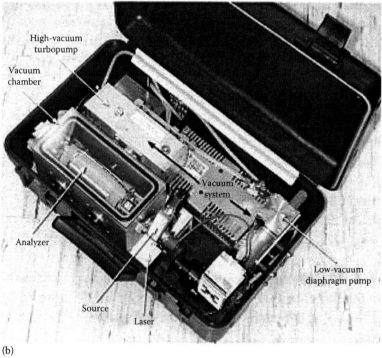

(b)

FIGURE 8.19 Suitcase TOF-MS system (a) in the breadboard configuration for testing and (b) packaged in a small Pelican case for demonstration purposes (data system not shown). (From Ecelberger, S.A. et al., *Johns Hopkins APL Tech. Digest*, 25(1), 14, 2004.)

8.8.3   Source/Analyzer

The heart of the Suitcase TOF is the source/analyzer. Conceptually, a linear TOF-MS (see Section 3.2.1) is very simple. Analyte and matrix ions formed in the electric field of the source region are all accelerated to the same KE. A flat sample plate held at a high voltage (≈ 10 kV) separated by a small distance from an extraction grid at ground potential typically defines a source. After exiting the source, the ions drift in a field-free region until they strike a detector. Because all the ions have the same KE, a more massive or heavier ion will have lower velocity than a less massive or lighter ion.

One method for correcting the initial energy distribution and decreasing the peak width is to use a "reflectron", a retarding electric field that reflects the ions back along their original path to strike a detector placed at the ions' spatial focus. The reflectron increases signal resolution by correcting for the initial energy spread and increasing the effective drift length of the analyzer. Finally, the analyzer uses a custom microchannel plate (MCP) detector assembly that reduces the signal ringing that is often associated with coaxial detectors.

8.8.4   Electronics/Data System

The Suitcase TOF uses a mix of commercial and custom electronics to control the instrument and collect the data. Five high-voltage power supply modules are used to manually control the source, reflectron, and detector voltages. A full-size LeCroy oscilloscope digitizes the TOF signal and averages between 10 and 100 individual shots to produce one TOF mass spectrum. A laptop computer running a National Instruments LabView program downloads the data from the oscilloscope and then processes, formats, stores, and presents the mass spectra. Efforts at greatly reducing the size, weight, power, and packaging of the electronics are planned to be addressed in the next prototype.

8.8.5   Performance

A series of biologicals ranging in mass from a few hundred daltons to over 60 kDa were run in parallel on the Suitcase TOF and commercial TOF-MS (Figure 8.20). According to authors of this study, the Suitcase TOF equaled the performance of the commercial TOF in resolution and sensitivity for some high mass compounds.

Overall, the investigators have demonstrated that a miniature man-portable TOF-MS can fullfill a number of needs. It is expected that enhanced optics, electronics, detectors, and other features will improve the performance of the Suitcase TOF, and continued reduction in size, weight, and power will improve its portability, leading eventually to a very powerful and simple-to-use battery-powered instrument (Ecelberger et al. 2004).

8.9   A TOF MINI-MS WITH AEROSOL COLLECTION, CAPTURE, AND LOAD-LOCK SYSTEM (BIOTOF)

The TOF mini-MS system (TOF-MMS) developed by researchers from JHU-APL is a MS-based analysis system for the rapid, unambiguous, in situ identification of a wide range of microbial organisms and toxic substances.

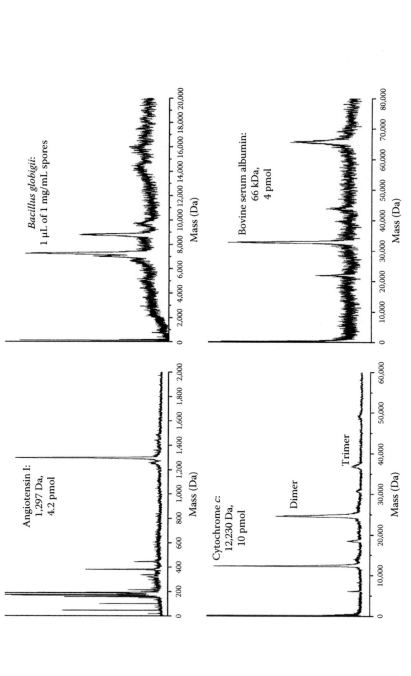

FIGURE 8.20 Representative spectra of calibration compounds covering the mass range of potential biomarkers of the biothreat agents. (From Ecelberger, S.A. et al., *Johns Hopkins APL Tech. Digest*, 25(1), 14, 2004.)

The major components of the system include an automatic aerosol collector, a sample processor, and a MS vacuum chamber load lock. A sample collection tape ties the components together by capturing samples from the collector, transporting the samples through a chemical processing stage, and introducing the samples into the MS detector. The tape replaces collection filters, processing slides, and insertion probes normally used to introduce samples into a spectrometer vacuum chamber (Figure 8.21). The BioTOF-MS features an air–air virtual impaction collection system that outputs a concentrated aerosol into a five-jet real impactor and subsequently deposits the particles onto a VHS cassette magnetic recording tape (Figure 8.22).

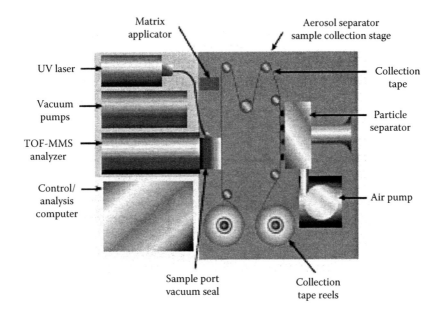

FIGURE 8.21 TOF-MS collection and analysis system.

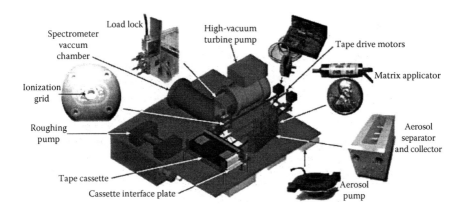

FIGURE 8.22 The assembly of TOF-MMS.

The aerosol sampling system performs the following tasks: (1) collects, concentrates, and separates aerosols from breathable ambient air at concentrations of 15 ACPs per liter of air and of 0.5–10.0 μm aerodynamic diameter, (2) captures infectious and toxic agents on a substrate in small spots that allow maximum coverage by an irradiating laser beam no larger than 1.0 mm in diameter, and (3) prepares the sample for the MALDI process by adding a matrix. This sample is introduced directly into the TOF-MMS, and furthermore is archived for post-analysis confirmation by storing processed samples and saving unprocessed samples from each collection period.

Since these activities are performed using a portable instrument, the process minimizes the time needed for collection and agent identification (not to exceed 5 min). Moreover, the instrument performs all functions automatically with no operator interaction. The instrument (1) has portability and reliability to survive vehicle transportation, (2) can be handled by two persons, and (3) can operate from a portable power source (Anderson et al. 1999).

The tape is transferred to a module for the deposition of the MALDI matrix onto the aerosolized sample spot. Sample treatment methods and incorporation of an internal standard are also conducted at this module. Following matrix application and drying, the sample is moved to the focal point of the laser, and upon laser firing, data are acquired from the detector. The data pass into a signal processing system that filters noise, converts the time signal to an m/z, determines a baseline, and identifies major peaks.

One can vary the amount of sample collected by adjusting the collection time and periodically advancing the tape. This allows adaptive processing of the spectra while sequentially adjusting the MALDI parameters for each simultaneously collected sample. In addition to serving as a sample collection medium, the tape may include a magnetic surface for electronic data recording. This creates a compact package for the collection and storage of the sample as well as data results, including collection location, time stamps, and analyses. The design also provides all the material needed for post-collection testing and verification.

8.10 MICROBELYNX™ A MALDI-TOF-MS SYSTEM FOR FINGERPRINT BACTERIAL ID

The developers of this system claim that their method allows the unique population of macromolecules expressed on the surface of bacteria to be rapidly sampled and characterized by molecular weight. The resulting mass spectrum provides a unique physicochemical fingerprint for the species tested that can be reliably matched against databases of quality-controlled reference mass spectra. They propose that this simple analytical method combined with suitable bioinformatics tools could be a powerful new tool for real-time detection and subtyping of bacteria.

Mass fingerprinting is rapid because the entire process from sample preparation to result takes only a few minutes for each test microorganism. Furthermore, sample preparation is quick and easy. In this implementation, intact cells from primary culture are smeared across a stainless steel target plate and allowed to co-crystallize with a UV-absorbing matrix. After drying, the target is placed in the MALDI-TOF-MS.

The microorganisms in the matrix are illuminated with a pulse from a nitrogen laser (337 nm). The resulting ionized macromolecules are mass analyzed and the results reported as a mass spectrum. The mass fingerprint of the test microorganism is then submitted to the MicrobeLynx™ search algorithm, which challenges an appropriately selected database from a range of quality-controlled bacterial reference mass spectra.

The current implementation assumes that the identification of bacteria could readily be achieved (e.g., from a small colony). It is a cost-effective approach because it requires minimal operator training and low per sample cost.

The database used for identification purposes has been developed in collaboration with The Molecular Identification Service Unit, Health Protection Agency, using validated strains from The National Collection of Type Cultures (NCTC), Colindale, London, UK.

Rapid identification of a microorganism from its characteristic mass fingerprint is easy with the MicrobeLynx pattern recognition Search Algorithm. MicrobeLynx applies a probabilistic mathematical strategy which compresses all the mass/intensity data in the test spectrum to a lower dimensional space, to give, in seconds, the identity of the bacterium with the best database match.

8.10.1   KEY FEATURES OF MICROBELYNX SEARCH ALGORITHM

- Rapid—takes a few seconds to complete a search
- Uses all the raw data in the test spectrum
- Results are simple to understand in MassLynx™ Browser format
- Comparative display of test spectrum and the best match
- Enables differentiation of closely related microorganisms
- Interactive review of all putative matches with the test spectrum
- The probability of the matches can be expressed as percentages

8.10.2   DATABASES

- Anaerobic Database
- Urinary Tract Infections Database
- Clinical Database
- *Bacillus* Database
- *Candida* Database
- Burkholderia Database

The absence of sample preparation coupled with rapid analysis and high throughput makes the technology attractive as a rapid microbial identification method. The method involves applying the bacterial cultures to the instrument plate wells, overlying the whole cells with a solvent matrix of CHCA for gram-negative bacteria and 5-chloro-2-mercaptobenzothiazole for gram-positive bacteria. The wells are then bombarded with a nitrogen laser that causes desorption of ionized cellular components that travel down a tube toward a detector. The time for the charged components to reach the detector operated in a positive ion detection mode using an acceleration voltage of +15 kV is a function of their KE (i.e., mass and charge). The detector signal is captured as a unique fingerprint for different species/stains of microorganism in the acquisition mass range of 500–10,000 Da. The MicroMass MALDI-TOF-MS-MicrobeLynx database currently has ~3500 spectral entries covering more than 100 genera and more than 400 different species. The time to process a sample is on the order of 3 min.

8.11   TIGER

Critical need exists for rapid, automated detection and identification of diverse and constantly evolving pathogens. In response to such needs, Ibis Therapeutics (Carlsbad, CA) developed the TIGER (Figure 8.23).

According to manufacturers, TIGER may identify quickly and precisely over 1400 known infectious pathogens (viruses, bacteria, fungi, and parasites) with new ones constantly evolving and the possibility of bioengineered ones being created, thus preventing widespread epidemics. Therefore, it could, in the national security arena, support biological weapons defense systems. Moreover, this system is characterized by a robust response and should be considered as an alternative to current procedures that involve individual time-consuming culture tests, which require a prior knowledge of the infectious disease and produce no results for pathogens that cannot be cultured.

TIGER provides the universal solution with rapid, robust, accurate identification for multiple and diverse pathogens in a single test:

- Rapid: Identification of organism type within hours and specific strain within 24 h
- Robust: Works on virtually all known, newly emergent, and bioengineered pathogens in a single test
- Culture free: No culture steps required
- Cost effective: Less expensive than running thousands of individual tests for detection
- Accurate: Thousands of samples tested with no false identifications
- Flexible: High throughput, multiple detection, and identification of diverse samples

Recently, Bruker Daltonics, a subsidiary of Bruker BioSciences Corporation (Billerica, MA), and Isis Pharmaceuticals entered into a manufacturing and distribution agreement for the Ibis T5000™ biosensor system. The Ibis T5000, developed by Isis' Ibis Biosciences division, is a universal biosensor system that can simultaneously identify thousands of types of infectious organisms in a sample (Figure 8.24). The Ibis Biosciences division has been funded by U.S. government agencies including the Defense Advanced Research Projects Agency (DARPA), the National Institute of Allergy and Infectious Diseases (NIAID), the Centers for Disease Control and Prevention (CDC), the Federal Bureau of Investigation (FBI), the Department of Homeland Security (DHS), and others.

Bruker Daltonics is the exclusive manufacturer of the Ibis T5000 biosensor system, which incorporates Bruker Daltonics' ESI-TOF-MS (micro-TOF). The Ibis T5000 utilizes the TIGER methodology, which is a combination of genomics, mathematical

The TIGER process

Step 1: Identify useful regions in the genome

Variable DNA sequences flandred by conserved sequences

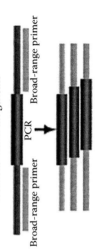

Conserved DNA | Variable DNA | Conserved DNA

Genomes have conserved regions that can be used to broadly amplify all forms of life. Some conserved regions surround highly variable regions whose sequences uniquely identify specific organisms. By looking at several of these regions (triangulation), TIGER separates strains and identifies emerging or engineered organisms.

Diverse sample

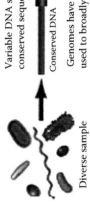

Step 2: Apply PCR amplification

Broad-range primers interrogate multiple locations within the genome

Broad-range primer

PCR

Broad-range primer

Step 3: Measure and analyze

– Mass spectrometry (electrospray ionization time-of-flight mass spectrometer) gives the mass of FCR products with high precision

– Triangulation—measures multiple signals from a threat agent

– Together, they provide a base composition fingerprint for all organisms in the sample

Step 4: Identification

Base composition provides the unique fingerprint of the agent

As–Y7
Gs–30
Cs–N4
Ts–6t

Precise number of As–Gs–Cs–Ts identifies the agent as known or unknown

FIGURE 8.23 The process used for TIGER.

FIGURE 8.24 Ibis T5000 Universal Biosensor. Key modules include amplicon purification and the MS. Precise molecular weight determination of amplicons yields unambiguous base compositions that are used to uniquely "fingerprint" each pathogen. The automated system is capable of analyzing over 1500 PCRs in 24 h.

modeling, MS, and molecular amplification to generate a "fingerprint" of each bacterium or virus, allowing it to identify virtually any bacteria or viruses present in a sample. In addition, the Ibis T5000 biosensor system can rapidly identify or classify organisms that are newly emerging, genetically altered, or unculturable. The Ibis T5000 biosensor works with many different types of infectious samples from human samples, such as throat swabs or sputum, to environmental samples such as soil or air.

This universal biosensor system requires the appropriate reagents (PCR primers, enzymes, etc.) and databases; however, it has the potential to meet diverse needs for the identification of infectious agents. The Ibis T5000 system can identify or classify and quantify a broad range of pathogens, and can also yield details on strain type. Ibis plans to commercialize the Ibis T5000 biosensor system and associated reagents to government customers for use in biowarfare defense, epidemiological surveillance, and forensics; and to nongovernment customers for use in pharmaceutical process control, hospital-associated infection control, and infectious disease diagnostics.

8.12   MICROORGANISM IDENTIFICATION SYSTEM BASED ON MALDI-TOF-MS BIOPROFILER

Bruker Daltonics, Inc. (Bellerica, MA) developed a microorganism identification system based on the spectral analysis of ionized cellular components obtained with their MALDI-TOF-MS instruments. They completed the analyses with BioProfiler software. In their approach, bacterial isolates are grown in a nutrient medium, harvested, and submitted for a MS/bioinformatics study. The spectra are obtained on a MALDI-TOF-MS (e.g., Bruker FLEX-series of laser desorption/ionization TOF-MS with instrument control software) by using a standard automatic acquisition in a linear positive ion mode in a mass range of 2–20 kDa and CHCA as a matrix. To

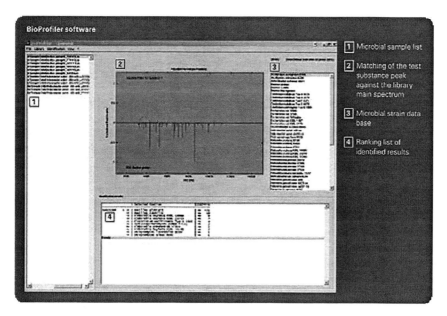

FIGURE 8.25 Screen capture of the BioProfiler software used for microorganism identification.

differentiate between the strains, statistically relevant data sets are subjected to PCA and mass spectra–based clustering. For microorganism identification purposes, spectra are processed using the MALDI-TOF-MS microbial spectra database and profile recognition/comparison algorithm BioProfiler.

A BioProfiler software tool (Figure 8.25) was developed for the reliable identification of unknown microorganisms from their protein fingerprint in the MALDI-TOF mass spectrum by comparing their individual peak lists to the available database. Based on a flexible database approach, the data obtained by MALDI-TOF-MS from cultivated bacterial colonies can be easily analyzed.

8.12.1 KEY FEATURES

- Ready-to-use databases are available including a database for bacterial BWAs
- Adjustable to new threats by creating your own database
- Training mode to extend existing databases or those provided by Bruker Daltonics
- Scoring factors allowing the interpretation of identification results

8.12.2 OPERATION

The BioProfiler software offers two levels of operation. Easy and fast identification incorporates loading the library and starting the pattern matching–based evaluation procedure. It incorporates the measurement of several substances in one go. Even different microbial strains of the same species can be distinguished with the described method. The advanced level enables laboratory professionals to create an individual data base out of their own collected available samples in order to address new threats.

8.13 HANDHELD RECTILINEAR ION TRAP MS (MINI-10)

Over the past decade, researchers in the laboratory of R. Graham Cooks (Chemistry Department, Purdue University, West Lafayette, IN) developed a series of miniature instruments based on the cylindrical ion trap (Patterson et al. 2002), and later on the rectilinear ion trap (RIT) geometry (Gao et al. 2006). The RIT geometry is very advantageous for miniaturized instruments because it allows for the large ion storage capacity characteristic for the linear ion trap and combines it with the design simplicity of the cylindrical ion trap.

Mini-10 is a shoebox-sized, handheld (10 kg) RIT-MS that is battery powered and consumes less than 75 W of energy. Nevertheless, Mini-10 gives unit mass resolution and an upper m/z limit of 600. Furthermore, it has MS/MS capabilities. Initially, the Purdue University team reported on the implementation of a membrane inlet with the traditional EI ion source for this instrument. However, recently they coupled AP ionization methods, ESI, and DESI to Mini-10 (Keil et al. 2007), thus making it more attractive for bio-applications. The Mini-10 is a fully autonomous MS/MS. Its basic characteristics are summarized in Table 8.3.

A schematic representation of the handheld MS with an ESI inlet is shown in Figure 8.26a. In this design, a narrow stainless steel capillary was used as the atmospheric inlet that transported ions into the ion trap due to different pressures between the ambient and the mass analyzer vacuum. A considerable challenge in coupling AP ionization methods to miniature MS systems is the reduced pumping capacity of a portable MS system, due to size and power consumption restrictions. A Mini-10 uses an oil-free two-stage diaphragm pump (5 L/min, KFN Neuberger) and a micro-turbopump (10 L/s; Pfeiffer TPD 011) and nickel–metal hydride batteries, which provide surge power capabilities that outperform lithium batteries. The control board can communicate with other computers via a TCP/IP port and allows the operator to control the Mini-10 through a computer running Microsoft Windows XP Professional. Moreover, it can be connected into a wired network and operation of multiple units can be controlled by a single remote computer.

TABLE 8.3
Characteristics of the Rectilinear Ion Trap MS—Mini-10

System Characteristics (m/z)	Performance Characteristics
Resolution—1	Linear dynamic range—3 orders of magnitude
Mass range—up to 600	Sensitivity—low ppb detection limit
Typical peak widths (FWHM): 3	*Physical Characteristics*
Ionization	Size: L—32 cm; W—22 cm; H—19 cm (0.013 m³)
Electrospray/desorption electrospray ionization	Mass—10 kg
Electron impact/glow discharge	Power 75 W (2–4 h of field battery operation)

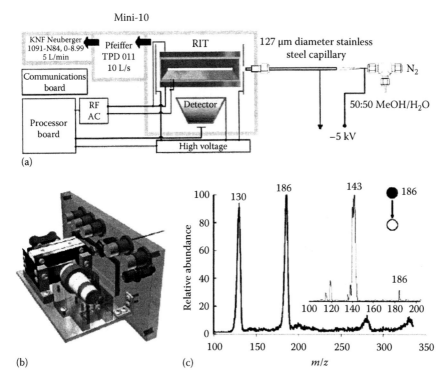

FIGURE 8.26 Handheld RIT-MS (Mini-10). (a) Schematic of the handheld MS with atmospheric inlet added and (b) a view of the components inside the manifold. (c) Typical ESI mass spectrum obtained of a 1/1 mixture of dibutylamine $(M+H)^+$, m/z 130 and tributylamine $(M+H)^+$, m/z 186. Inset: MS/MS spectrum of tributylamine. (From Keil, A. et al., *Anal. Chem.*, 79(20), 7734, 2007.)

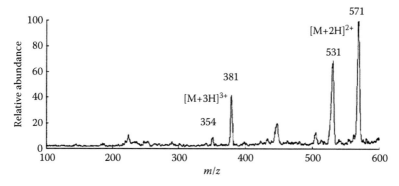

FIGURE 8.27 ESI mass spectrum of an equimolar mixture of bradykinin and the synthetic peptide KGAILKGAILR. (From Keil, A. et al., *Anal. Chem.*, 79(20), 7734, 2007.)

Unfortunately, supplementary components of the ionization methods are still to be miniaturized and capabilities for detection of bioagents have to be demonstrated. The potential for using this instrument in proteomics-based applications is demonstrated in Figure 8.27. In this figure, capabilities to acquire the ESI mass spectrum of peptides is demonstrated by electrospraying an equimolar mixture of bradykinin and the synthetic peptide KGAILKGAILR (Keil et al. 2007). Furthermore, a polymer-based RIT mass analyzer was fabricated by using stereolithography, which is light, small, less expensive and with lower power consumption requirements (Fico et al. 2007). Therefore, given the level of performance observed for a Mini-10, the RIT-based instruments seem to be good candidates for fieldable MS connected into a network; however, further development is still required.

9 Current and Future Trends in Using MS for Microbial Detection and Identification

Charles H. Wick

CONTENTS

9.1 REQUIREMENTS FOR MICROBIAL DETECTION AND IDENTIFICATION

The basic requirement is acceptable accuracy for all classes of microbes. Anything short of this basic need is a work-around of the real requirement. How we obtain this capability depends on three parts: (1) collecting and processing the sample, (2) analyzing the sample through something like a MS instrument that simply collects "all" the information from the sample, and (3) software to analyze.

First, responders need these three systems to be able to function accurately and frequently under less than optimum conditions. It would be nice if the field system was integrated into a single portable instrument with disposable consumables such as sample collectors. This needs changes and becomes less rigorous for vehicle-based and laboratory-based systems.

These basic requirements are not withstanding; it is also imperative that this new capability has its own operational requirements and should not be tossed in as some subsystem to a larger system. Often those general requirements for the main system are not mutually consistent with microbe detection and identification and frequently do not specify false negatives or false positives for either system.

9.1.1   First-Tier Sensors

Currently, new MS-based platforms are emerging as molecular-level technologies for rapid detection and identification of potential biothreats. Some of these methods could be used in reconnaissance vehicles or in field deployable laboratories, while the others are more suitable for diagnostic or forensic laboratories for processing samples with the highest scrutiny and to validate findings based on other assays. Therefore, there is a need for guidelines that would take into account new capabilities of MS technologies.

Some of these MS-based sensors have been developed and tested. Among them, a single-particle analyzer termed BAMS has been developed by the Lawrence National Laboratory (Section 8.6) and a fieldable biodetection system that uses MALDI-aerosol-TOF-MS (Section 8.5) has been developed in The Netherlands. Other advances in portable MS instrumentation that have been coupled with ambient ionization techniques have the potential to greatly enhance in situ analysis. For example, DESI has been coupled with mobile or even handheld instruments (Keil et al. 2007), thus the first step was taken toward a handheld miniature MS capable of AP ionization. Further developments of the MS technique can be expected to significantly improve the detection of bioagents in the field. For instance, recently Shin et al. (2007) proved that a robust method based on DESI-MS is capable of a rapid and accurate molecular mass determination for proteins up to 17 kDa. Moreover, the same group demonstrated detection of the bacteriophage MS2 capsid protein from crude samples with minimal sample preparation by DESI-MS. Software advances (Deshpande et al. 2011) further facilitate the analysis of MS outputs and are frequently portable to many different MS platforms. This convergence of the three components can be expected to produce a first-tier capability with acceptable accuracy to satisfy the basic requirement of acceptable accuracy for all classes of microbes.

9.1.2   Second-Tier Sensors

The detection of a microbe of interest by a "first-tier" sensor should be confirmed by a second-tier sensor. Typically, this conformation sensor should be capable of providing (a) higher specificity (orthogonal method), (b) speed (e.g., less than an hour), (c) high throughput, (d) automation that is applicable to analysis of a wide range of threats and their mixtures, and (e) low consumable cost and naturally satisfy the basic requirement of acceptable accuracy for all classes of microbes.

Fingerprint-type approaches are quite successful in laboratory settings to discriminate bacteria up to the strain level and a considerable progress in this area has been achieved, especially in case of fingerprinting bacteria by using MALDI-MS technology. However, it should be noted that any fingerprinting method requires pure isolates to be accurate; therefore, the presence of naturally occurring microorganisms and other background components makes such an approach highly questionable for environmental monitoring. Nevertheless, these methods could be very valuable for monitoring the decontamination process, or for rapid typing and subtyping of pure isolates.

Sequence-based detection represents a more refined use of MS technology for highly specific detection and identification of microorganisms. Therefore, MS/MS techniques are excellent candidates for fulfilling second-tier sensor requirements because these platforms are capable of rapidly confirming the initial biothreat detection obtained by other sensors like simple MS sensors or other technologies (PCR, immunoassays) with high specificity. These MS/MS methods rely on soft ionization techniques, such as MALDI or ESI, and are capable of providing genome traceable sequencing information (Chapter 7). Furthermore, MS/MS-based methods can be viewed as orthogonal biothreat detection and identification systems in comparison to PCR and antibody-based approaches. Overall, tandem MS-based methods fulfill all the requirements for a "second-tier" system until they become portable and a useful "first-tier" device. One very convenient feature of a MS-based platform is that the files can be sent via electronic means to another MS platform for conformation.

9.2 INTEGRATION AND AUTOMATION OF SAMPLE COLLECTION, PROCESSING, AND ANALYSIS

The key to developing rapid and reliable methods for the sensitive, and near real-time detection of pathogens in environmental samples lies in the ability to integrate and automate sample collection, processing, and concentration of target analytes with a MS platform. Microorganisms can be concentrated and separated from other matrix components in a number of ways described in Chapter 5. However, the overwhelming majority of published procedures are not suitable for field applications. For example, antibodies coupled to magnetic beads could be used to separate organisms from environmental constituents such as water or other fluids, and these approaches are widely used in many laboratory applications. However, novel, more robust methods are needed that do not use antibodies and are capable of providing at least semispecific capturing of targeted microorganisms.

9.2.1 COLLECTOR-FREE APPROACHES

Currently, there are MS-based first-tier sensors available for analyzing an aerosol stream directly without the use of a collector. For example, using a BAMS fluorescent aerodynamic particle sizer (FLAPS) system, aerosol at a flow rate of 1 L/min is passed through a detector that measures single-particle aerodynamic size and fluorescence before particle analysis. In another example, a MALDI-MS has been fitted with a continuous flow sample conditioning system that applies the matrix coating without first collecting the particulate matter. The sampled aerosol is passed through an evaporation and condensation system, where the matrix is condensed onto the particulate matter, and the coated aerosol subsequently drawn into the mass spectrometer.

9.2.2 SAMPLE COLLECTION

The confirmatory analysis of initial bioagent detection by MS platforms usually requires samples in the liquid state. Therefore, sample collection approaches use concentrators (Section 4.2) that must efficiently transfer the aerosol into the hydrosol state. Nevertheless, there is always a concern that it is difficult to find the target of interest when the background microflora can be many orders of magnitude greater than the pathogen. Therefore, additional purification may be required under certain circumstances.

Although there is a lack of fully integrated systems for MS-based analysis in the field, there are numerous examples of integrated systems, e.g., based on immunomagnetic cell capture, that have a completely automated procedure for sample collection and target analyte purification from its background. These methods demonstrate that the needed sensitivity and specificity boost for extremely low concentrations of a pathogen is an achievable goal for MS-based applications. In addition, broader spectrum approaches such as lectin and carbohydrate capture, and other semi-specific capturing surfaces could be used as well. When moving to these broader spectrum capture approaches, evaluation of the ability of the MS-based detection system to find target organisms in high background microflora will need to be investigated.

There are many commercially available kits and reagents that target the extraction and separation of community DNA or RNA from such environmental samples as soils, sediments, food, and water for use in PCR and/or nucleic acid hybridization assays in a semiautomated manner (Bruckner-Lea et al. 2002). Many elements of these approaches could be easily adopted for sample preparation protocols suitable for MS-based assays. For example, the Pacific Northwest National Laboratory (PNNL) sample preparation approach utilizes microparticles with surface chemistries that are designed for capturing the analyte of interest. The central feature of the PNNL Biodetection Enabling Analyte Delivery System (BEADS) platform is a renewable surface separation column. Figure 9.1 illustrates the two types of columns: (a) a rotating rod column and (b) a magnetic column that use microbeads and supermagnetic particles, which are automatically packed into the renewable surface chamber. The surface chemistry of the particles is designed to capture the cells or biomolecules of interest. In short, environmental samples flow through such columns and target analytes are trapped on the microparticle surfaces, while the remainder of the sample flows through columns. In this way, microorganisms or their constituents (e.g., proteins, DNA, RNA) are concentrated and purified.

At the end of the process, the microparticles are flushed from the column, automated routines thoroughly clean the system to remove any potential contaminants from the previous sample, and then the column is renewed with fresh microparticles. The BEADS platform provides a highly flexible method for processing environmental samples to detect pathogens. The components are completely interchangeable, allowing the user to switch between rotating rod and magnetic renewable surface columns, depending on the type of microparticles used for the sample preparation task. The integrated system is entirely computer controlled and includes a graphical

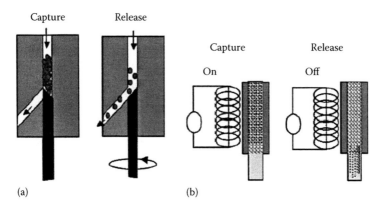

FIGURE 9.1   Renewable surface microcolumns used for cell capture and nucleic acid purification. (a) Rotating rod microcolumn is used for automatically capturing and releasing microparticles ranging from 5 to 150 μm in size. When beads are captured, fluid can still pass through the device, allowing perfusion of the beads with samples and reagents. (b) The magnetic column includes a nickel foam core to capture and release magnetic particles throughout the flow path, using an electromagnet. (From Bruckner-Lea, C.J. et al., *Anal. Chim. Acta*, 469(1), 129, 2002.)

user interface for changing process parameters depending on the sample-processing needs. The system is designed for PCR analysis for trace detection pathogens in environmental samples, and includes automated whole cell capture onto immuno-magnetic beads and flow-through PCR. However, the process could be easily modified for a downstream detection by MS methods. For instance, the MS platform could be coupled in-line with the BEADS system for base compositional analysis of PCR amplicons. Furthermore, by using a suitable lysis cocktail, selectively captured cells could be raptured and other released target constituents, e.g., RNAs or proteins, selectively captured and submitted for analysis by a MS-based platform.

One of the emerging technologies for rapid bacteria identification uses MS-based sequencing of protein amino acids as fast readout of genomic information that can be used both for detection and identification purposes. This would indicate that this approach is rapid, safe, and automated sample processing is required for timely detection and identification of biological pathogens using this approach. For example, recently, microfluidic microbial sample-processing modules that could be integrated with the MS systems for a proteomics-based approach have been developed for analysis of aerosols by ECBC (Figure 9.2). Such modules should pose capabilities to break apart biological agents collected in a fluid by an aerosol concentrator; to separate, purify, and concentrate the protein portion of a sample; and to digest proteins into small peptides with a proteolytic enzyme.

It is anticipated that future developments will focus on the integration of sample processing in the microfluidic format. For example, Wallman et al. (2006) developed a capillary filling microfluidic proteomic sample-processing system in an automated setup. Built-in force feedback control ensures a precise and robust microchip docking/handling in three dimensions. The system throughput ranges typically from 50 to 100 samples/h.

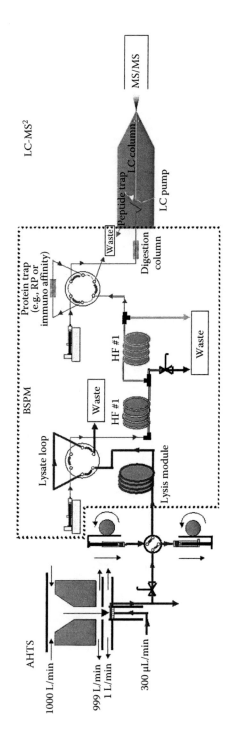

FIGURE 9.2 Schematic representation of the sample-processing system based on aerosol pre-concentration and transfer to the hydrosol phase (AHTS), followed by lysis, extraction, and digestion of proteins to peptides (biological sample-processing module [BSPM]) and proteomic analysis using LC-MS/MS platform (LC-MS²).

9.3 SENSITIVITY AND SPECIFICITY CONSIDERATIONS

9.3.1 PROTEOMICS

There is a need for greater sensitivity to better detect, identify, and quantify biomolecules derived from microorganisms. The real limitations to achieving sufficient depth with present proteome measurements are essentially derived from considerations associated with the amount of sample available and the sensitivity of the measurement method. In the case of potential field applications, measurement sensitivity becomes the limiting factor. The overall sensitivity of ESI-MS is limited by factors that include both ionization efficiency and ion transmission efficiency into and through the MS analyzer. At the liquid flow rates of conventional LC separations, ESI-MS response typically appears concentration-sensitive rather than mass-sensitive. Increasing the flow rate does not greatly increase the signal. However, as flow rates are lowered, the smaller charged droplets generated by an electrospray result in increased ionization efficiency (i.e., transfer of an ion from solution to gas phase) and also permit the ESI emitter to be positioned closer to the MS inlet to allow more efficient transport to the MS analyzer, both of which provide increased sensitivity. For example, ESI-MS analyses that used flow rates of ~20 nL/min demonstrated significantly increased sensitivity compared with flow rates typically applied with LC separations (Schmidt et al. 2003). Furthermore, Quanzhou et al. (2006) reported that the use of 10 μm i.d. silica-based monolithic LC columns providing flow rates on the order of 10 nL/min (i.e., nano-flow) increased sensitivity and improved quantitation.

Another improvement has been derived from the use of automated gain control (AGC) with ion trapping instrumentation, which constrains ion populations in ion traps to levels that maintain the desired mass measurement accuracy. Since the present maximum ion transmission rates from an ESI source ($>10^8$ charge/s) can exceed trap capacity by >100-fold, a potentially useful signal is often wasted. Application of AGC extends the overall dynamic range by more than 10-fold, which in turn provides large increases in detection and identification capabilities. Such developments will continue to expand the role of ESI-MS in biological research and, increasingly, extend its use in biothreat agent detection applications (Table 9.1).

The currently available low-resolution laboratory systems (e.g., ion trap MS) can provide sensitivity for sequence-based peptide identification at the level approaching 1 fmol. At this sensitivity level, about 3000 cells would be required for successful identification of bacteria based on sequencing information derived from the most expressed proteins (Table 9.2). For example, there are about 20,000 ribosomes per each *E. coli* cell and each ribosome is composed of 55 proteins; therefore, each of the

---

TABLE 9.1

Detection and Identification Performance Objectives for CBMS

CWA liquids on ground	Identify at <0.4 mg/m² in <45 s
CWA vapors in the air	Identify at <0.1 mg/m³ (blister), <0.01 mg/m³ (nerve) in <45 s
BWA in air	<25 agent-containing particles/L of air (ACPLA) in <5 min; "better than current systems"

TABLE 9.2
Number of Cells Required for Detection and Identification of Bacteria Related to Given Protein Present in Cell at Indicated Copy Number and Sensitivity of MS

Protein Copies/Cell	Number of Cells Required/MS Sensitivity	
	10 fmol	1 fmol
10	6.02×10^8	6.02×10^7
100	6.02×10^7	6.02×10^6
1,000	6.02×10^6	6.02×10^5
10,000	6.02×10^5	6.02×10^4
100,000	6.02×10^4	6.02×10^3
1,000,000	6.02×10^3	6.02×10^2

ribosomal proteins is present at the same copy number. When an instrument operates at the sensitivity level of 1 fmol, it means that about 30,000 cells are needed for the successful sequencing of proteins and the information may be used for bacteria identification. However, in the case of protein chain elongation factors (e.g., Tu), each cell may contain about 10 molecules of Tu per ribosome; hence, by using sequences derived from this highly expressed protein (about 200,000 copies per cell), identification may require only 3000 cells (Table 9.2).

9.3.2 Ion Mobility Spectrometry

Recent developments in instrumentation for ion mobility spectrometry (IMS), a gas-phase post-ionization separation method, coupled with MS (IM-MS) reveal potential applications of this relatively new technique for rapid, high-resolution separations of analytes based on structure (ion conformation) and *m/z* ratios. Although IM-MS separations can be achieved at resolutions comparable to HPLC and capillary electrophoresis (CE), most IM-MS instruments are operated at a relatively low IM resolution to reduce analysis time, maximize sample throughput, and reduce the complexity of interfacing IM with MS. Proteomic-scale identification of microorganisms requires a large number of measurements (determination of molecular weight and amino acid sequence) and broad dynamic range; therefore, such techniques must afford high throughput at low limits of detection, high sensitivity, and wide dynamic range. Current efforts are motivated by the unique capabilities afforded by IM-MS and IM-MS/MS such as suppression or elimination of chemical noise, rapid (μs–ms) separation of complex biological mixtures, and nearly simultaneous sequencing of peptides (McLean et al. 2005).

Several instrument configurations could be applied to the detection and identification of biothreat agents. Because there is a balance between ion transmission (sensitivity) and mass spectral resolution that can be manipulated, lowering of the resolution of the mass measurement enhances limits of detection and sensitivity relative to high-resolution mass measurement experiments, because all of the analyte signal is contained in a single channel rather than being dispersed across the entire isotope cluster. For example, for proteomics-based identification of bioagents, it is often desirable to observe a larger number of low-resolution peptide fragment ion signals in a MS/MS experiment, rather than a lower number of high-resolution peptide isotope cluster ion signals. However, mass measurement accuracy on peptide ion signals where the isotope cluster is not resolved is limited to ca. 50 ppm and may be insufficient for high confidence level protein or oligonucleotide (e.g., PCR amplicon) identification. However, the resolutions of both the IM and MS dimensions could be tailored to provide an optimal balance between sensitivity and resolution for high confidence level protein identification that could be used in the case of protein toxin analysis. Generally, instrumental advances are aimed at providing higher sensitivity, faster analysis times, and higher information content. Both sensitivity and dynamic range can be significantly improved in IM-MS by multiplexing ion injection into the drift cell (MALDI or ESI) that allows to increase throughput by factors of 10^2 to 10^4 (McLean et al. 2005).

9.4   APPLICATIONS

With the use of MS platforms and advanced software, it is now possible to complete detection and identification of all classes of microbes in a minute rather than days and to determine classification by computer rather than tedious laboratory work. This capability also allows for other interesting applications.

The regular uses of detection and identification of microbes in routine such as in infectious diseases, environmental situations, veterinary situations, and agriculture are all indicated. Just to know if a patient has a bacterial component or not would allow a more prudent use of antibiotics. Animal diseases can be monitored and detected. New strains of microbes can be studied and, most important, the detection and early identification of emerging diseases can be rapidly analyzed.

Regarding the issue of emerging diseases, one feature of a MS approach is the ability to have samples collected and processed by MS worldwide and then have the raw files electronically sent to others for conformation and further study. This ability further allows for a centralized MS platform to receive and process all the files in the network for coordination and response. It would be possible for this centralized component to monitor changes occurring in the genome and provide for an accurate analysis to recommend vaccine approaches.

9.5   CONCLUDING REMARKS

MS has emerged as an indispensable tool for the biological sciences as a result of developments that took place during last two decades. They include the emergence of novel ionization methods, mass analyzers, and sensitive ion detection methods, which are circumscribed in this book. These achievements allow for detection, quantification, and in-depth structural analysis of many types of biomolecules that could be utilized for sensitive and highly specific detection and identification of microbial agents. These include biopolymers like proteins, nucleic acids, and complex lipids, in addition to traditional applications of MS for analysis of small molecules. Furthermore, in many cases only a limited sample preprocessing is required before submitting a sample for MS analysis.

One can expect that future developments will include both miniaturization of MS and the enhancements in ambient pressure ionization methods, combined with more efficient ion manipulation in multistage mass analyzers. Therefore, it is likely that also measurements suitable for obtaining protein sequence information, for instance, could soon be made possible. Such a system could be used to analyze air samples for rapid identification of pathogens as well as allowing for a reliable monitoring of a decontamination process.

Although profiling of bacterial proteins and other cellular components using MALDI-MS or DESI provides descriptive characteristics suitable for fast discrimination and typing of microorganisms, this approach is not suitable for direct analysis of bioagents from complex environmental matrices. Therefore, reliable detection and identification of biothreat agents will benefit from technologies that provide rapid and preferably genome-based data for the classification and identification of pathogenic and nonpathogenic strains. For instance, molecular approaches use peptide ions derived from microbial proteins, which are fragmented by CID or during TOF post-source decay, to reveal protein sequence information. This information can be used for the detection and confirmatory identification of bioagents by searching protein databases. Currently, it seems that "shotgun" digestion of bacterial proteomes combined with LC-MS/MS analysis of the generated peptides may provide sequence information that is sufficient even for the exploratory classification of complex environmental samples and is suitable for phylogenetic classification of microbes. For example, assignments of identified experimental peptide sequences to database proteomes/genomes in the database create phylogenetic profiles of the peptides. These profiles may be analyzed using numerical taxonomy methods to reveal groupings of an investigated strain with database bacteria that are suitable to infer a taxonomic position and confirm identity of bioagent.

MS-based proteomic methods can be used for pathogen detection, classification, and identification through approaches that complement DNA-based assays and provide orthogonal detection capabilities to prevent system-wide false positives or negatives. Moreover, because protein sequences are more conserved than DNA sequences, the investigation of proteomes can provide a clearer picture of functional relatedness by eliminating interclonal DNA divergence that is nonessential for the functional or pathogenic perspective. In addition, proteomics-based methods may facilitate the detection of virulence proteins and those virulence proteins that were expressed from genes deliberately modified. The latter can include, for example, production of alternative codons for some amino acids to avoid nucleotide detection.

During the next 5 years, the number of fully sequenced bacterial genomes will approach and likely exceed the number of known bacterial genera (1194) because currently more than 725 full genome-sequencing projects are in progress. Although some genera and species will still be underrepresented in the database, it is clear that the taxa most important from the pathological and environmental standpoint will be represented by many strains, thus assuring a solid foundation for a growing use of proteomics methods for detection and identification and diagnostic of microbes. The combination of these unprecedented resources with the expected progress in (a) automated sample preparation, (b) peptide separation techniques utilizing microfluidic devices, (c) novel MS instrumentation, and (d) bioinformatics methods will allow for development of a robust platform for fast, efficient, and a comprehensive, comparative, proteogenomic analysis of bacteria during a few minutes time frame. Thus, the expected progress in experimental and computational approaches combined with a sufficient knowledge base will create an environment to perform a successful identification process for the majority of cultivable bacteria. This progress should also provide information suitable to predict important biological properties such as pathological potential or disease outcomes from human and animal pathogens.

References

Aebersold, R. 2003. A mass spectrometric journey into protein and proteome research. *Journal of the American Society for Mass Spectrometry* 14(7):685–695.

Aebersold, R. and D. R. Goodlett. 2001. Mass spectrometry in proteomics. *Chemical Reviews* 101:269–296.

Aebersold, R. and M. Mann. 2003. Mass spectrometry-based proteomics. *Nature* 422(6928): 198–207.

Akoto, L., R. Pel, H. Irth, U. A. T. Brinkman, and R. J. J. Vreuls. 2005. Automated GC-MS analysis of raw biological samples-application to fatty acid profiling of aquatic micro-organisms. *Journal of Analytical and Applied Pyrolysis* 73(1):69–75.

Alfonso, C. and C. Fenselau. 2003. Use of bioactive glass slides for matrix-assisted laser desorption/ionization analysis: Application to microorganisms. *Analytical Chemistry* 75(3):694–697.

Amado, F. M. L., P. Domingues, M. G. Santana-Marques et al. 1997. Discrimination effects and sensitivity variations in matrix-assisted laser desorption/ionization. *Rapid Communications in Mass Spectrometry* 11(12):1437–1452.

Amati, G., O. Belenkiy, B. Dassa, A. Shainskaya, and S. Pietrowski. 2003. Distribution and function of new bacterial intein-like protein domains. *Molecular Microbiology* 47(1):61–73.

Anderson, C. A. and M. A. Carlson. 1999. A time-of-flight mini-mass spectrometer: Aerosol collection, capture, and load-lock system. *Johns Hopkins APL Technical Digest* 20(3):352–362.

Anderson, D., I. Brookhouse, and K. Ralphson. 1999. Numerical identification and classification of microorganisms using MALDI-TOF. Paper presented at the *47th Annual ASMS Conference Mass Spectrometry and Allied Topics*, Dallas, TX, June 13–17, 1999.

Anderson, D. and I. J. East. 2008. The latest buzz about colony collapse disorder. *Science* 319(5864):724–725.

Anonymous. 2000. Analysis Informativo (Laboratory analysis of bees from Guadalajara, Spain. Servicio de Investigacion Agraria, Bee Pathology Laboratory, Central Apicola Regional, Marchamalo, Spain. Provided by M. Higes on December 13, 2000.

Antoine, M. D., M. A. Carlson, W. R. Drummond et al. 2004. Mass spectral analysis of biological agents using the bioTOF mass spectrometer. *Johns Hopkins APL Technical Digest* 25(1):20–26.

AOAC International. 2004. Initiative yields effective methods for anthrax detection; RAMP and MIDI, Inc., Methods approved. *Inside Laboratory Management* 10(3).

Arnold, R. J., J. A. Karty, A. D. Ellington, and J. P. Reilly. 1999. Monitoring the growth of a bacteria culture by MALDI-MS of whole cells. *Analytical Chemistry* 71(10):1990–1996.

Arnold, R. J. and J. P. Reilly. 1998. Fingerprint matching of *Escherichia coli* strains with matrix-assisted laser desorption/ionization time-of-flight mass spectrometry of whole cells using a modified correlation approach. *Rapid Communications in Mass Spectrometry* 12(10):630–636.

Arnold, R. J. and J. P. Reilly. 1999. Observation of *Escherichia coli* ribosomal proteins and their posttranslational modifications by mass spectrometry. *Analytical Biochemistry* 269(1):105–112.

Bailey, L. and B. V. Ball. 1978. *Apis* iridescent virus and "clustering disease" of *Apis cerana*. *Journal of Invertebrate Pathology* 31(3):368–371.

Bailey, L., B. V. Ball, and R. D. Woods. 1976. An iridovirus from bees. *Journal of General Virology* 31:459–461.

Baillie, L. W., M. N. Jones, P. C. Turnbull, and R. J. Manchee. 1995. Evaluation of the BIOLOG® system for the identification of *Bacillus anthracis. Letters in Applied Microbiology* 20(4):209–211.

Baker, A. C. and C. Schroeder. 2008. The use of RNA-dependent RNA polymerase for the taxonomic assignment of picorna-like viruses (order Picornavirales) infecting *Apis mellifera* L. populations. *Virology Journal* 5(10).

Bakhtiar, R. and Z. Guan. 2005. Electron capture dissociation mass spectrometry in characterization of post-translational modifications. *Biochemical and Biophysical Research Communications* 334(1):1–8.

Barinaga, C. J., D. W. Koppenaal, and S. A. McLuckey. 1994. Ion-trap mass spectrometry with an inductively coupled plasma source. *Rapid Communications in Mass Spectrometry* 8(1):71–76.

Barshick, S. A., D. A. Wolf, and A. A. Vass. 1999. Differentiation of microorganisms based on pyrolysis-ion trap mass spectrometry using chemical ionization. *Analytical Chemistry* 71(3):633–641.

Becnel, J. J. and S. E. White. 2007. Mosquito pathogenic viruses: The last 20 years. *Journal of the American Mosquito Control Assocation* 23(supplement 2):36–49.

Belgrader, P., D. Hansford, G. T. A. Kovacs et al. 1999. A minisonicator to rapidly disrupt bacterial spores for DNA analysis. *Analytical Chemistry* 71(19):4232–4236.

Bell, C. A., J. R. Uhl, T. L. Hadfield et al. 2002. Detection of *Bacillus anthracis* DNA by LightCycler PCR. *Journal of Clinical Microbiology* 40(8):2897–2902.

Benoit, P. W. and D. W. Donahue. 2003. Methods for rapid separation and concentration of bacteria in food that bypass time-consuming cultural enrichment. *Journal of Food Protection* 66(10):1935–1948.

Benz, I. and M. A. Schmidt. 2002. Never say never again: Protein glycosylation in pathogenic bacteria. *Molecular Microbiology* 45(2):267–276.

Bern, M., D. Goldberg, W. H. McDonald, and J. R. Yates, III. 2004. Automatic quality assessment of peptide tandem mass spectra. *Bioinformatics* 20(Supplement 1):i49–i54.

Bertrand, M. J., P. Martin, and O. Peraldi. 2000. *A New Concept in Benchtop Mass Spectrometer, MAB-ToF*. New Orleans, LA: PittCon.

Bilimoria, S. L. 2001. Use of viral proteins for controlling the boll weevil and other insect pests. *U.S. Patent* 6,200,561.

Birmingham, J., P. Demirev, Y.-P. Ho, J. Thomas, W. Bryden, and C. Fenselau. 1999. Corona plasma discharge for rapid analysis of microorganisms by mass spectrometry. *Rapid Communications in Mass Spectrometry* 13(7):604–606.

Black, G. E., A. Fox, K. Fox et al. 1994. Electrospray tandem mass spectrometry for analysis of native muramic acid in whole bacterial cell hydrolyzates. *Analytical Chemistry* 66(23):4171–4176.

Bligh, E. G. and W. J. Dyer. 1959. A rapid method of total lipid extraction and purification. *Canadian Journal of Biochemistry and Physiology* 37(8):911–917.

Bolbach, G. 2005. Matrix-assisted laser desorption/ionization analysis of non-covalent complexes: Fundamentals and applications. *Current Pharmaceutical Design* 11(20):2535–2557.

Bornsen, K. O., M. A. S. Gass, G. J. M. Bruin et al. 1997. Influence of solvents and detergents on matrix-assisted laser desorption/ionization mass spectrometry measurements of proteins and oligonucleotides. *Rapid Communications in Mass Spectrometry* 11(6):603–609.

Bothner, B. and G. Siuzdak. 1995. Electrospray ionization of a whole virus: Analyzing mass, structure, and viability. *ChemBioChem* 5:258–260.

Braun, V., M. Mehlig, M. Moos et al. 2000. A chimeric ribozyme in Clostridium difficile combines features of group I introns and insertion elements. *Molecular Microbiology* 36(6):1447–1459.

Breaker, R. R. 2004. Natural and engineered nucleic acids as tools to explore biology. *Nature* 432:838–845.

Brigati, J., D. D. Williams, I. B. Sorokulova et al. 2004. Diagnostic probes for *Bacillus anthracis* spores selected from a landscape phage library. *Clinical Chemistry* 50(10):1899–1906.

Bright, J. J., M. A. Claydon, M. Suofian, and D. B. Gordon. 2002. Rapid typing of bacteria using matrix assisted laser desorption ionization time-of-flight mass spectrometry and pattern recognition software. *Journal of Microbiological Methods* 48(2–3):127–138.

Bromenshenk, J. J. 2010. Colony collapse disorder (CCD) is alive and well. *Bee Culture* 138:51.

Bromenshenk, J. J., C. B. Hendeson, C. H. Wick et al. 2010. Iridovirus and microsporidian linked to honey bee colony decline. *PLoS One* 5(10):e13181.

Brown, K. 2004. Up in the air. *Science* 305:1228–1229.

Brown, S. D., M. R. Thompson, N. C. VerBerkmoes et al. 2006. Molecular dynamics of the *Shewanella oneidensis* response to chromate stress. *Molecular and Cellular Proteomics* 5:1054–1071.

Bruce, W. A., D. L. Anderson, N. W. Calderone, and H. Shimanuki. 1995. A survey for Kashmir bee virus in honey bee colonies in United States. *American Bee Journal* 135:352–355.

Bruckner-Lea, C. J., T. Tsukuda, B. Dockendorff et al. 2002. Renewable microcolumns for automated DNA purification and flow through amplification from sediment samples through polymerase chain reaction. *Analytical Chimica Acta* 469(1):129–140.

Bruins, A. P. 1991. Liquid chromatography-mass spectrometry with ionspray and electrospray interfaces in pharmaceutical and biomedical research. *Journal of Chromatography* 554(1–2):39–46.

Bruno, J. G. 1999. In vitro selection of DNA aptamers to anthrax spores with electrochemiluminescence detection. *Biosensors and Bioelectronics* 14(5):457–464.

Bruno, J. G. and J. L. Kiel. 2002. Use of magnetic beads in selection and detection of biotoxin aptamers by electrochemiluminescence and enzymatic methods. *BioTechniques* 32:178–183.

Bukhari, Z., R. M. McCuin, C. R. Fricker, and J. L. Clancy. 1998. Immunomagnetic separation of *Cryptosporidium parvum* from source water samples of various turbidities. *Applied and Environmental Microbiology* 64(11):4495–4499.

Bundy, J. L. and C. Fenselau. 1999. Lectin-based affinity capture for MALDI-MS analysis of bacteria. *Analytical Chemistry* 71(7):1460–1463.

Bundy, J. L. and C. Fenselau. 2001. Lectin and carbohydrate affinity capture surfaces for mass spectrometric analysis of microorganisms. *Analytical Chemistry* 73(4):751–757.

Burke, S. A., J. D. Wright, M. K. Robinson, B. V. Bronk, and R. L. Warren. 2004. Detection of molecular diversity in *Bacillus atrophaeus* by amplified fragment length polymorphism analysis. *Applied and Environmental Microbiology* 70(5):2786–2790.

Buttner, M. P., K. Willeke, and S. Grinshpun. 2002. Sampling and analysis of airborne microorganisms. In *Manual of Environmental Microbiology*, 2nd edn., C. J. Hurst, R. L. Crawford, G. Knudsen, M. McInerney, and L. D. Stetzenbach, eds. Washington, DC: ASM Press, pp. 814–826.

Cain, T. C., D. M. Lubman, and W. J. Weber, Jr. 1994. Differentiation of bacteria using protein profiles from matrix-assisted laser desorption/ionization time-of-flight mass spectrometry. *Rapid Communications in Mass Spectrometry* 8(12):1026–1030.

Camazine, S and T. P. Liu. 1998. A putative iridovirus from the honey bee mite, Varroa jacobsoni Oudemans. *Journal of Invertebrate Pathology* 71(2):177–178.

Campitelli, L., I. Donatelli, E. Foni et al. 1997. Continued evolution of H1N1 and H3N2 influenza viruses in pigs in Italy. *Virology* 232(2):310–318.

Castanha, E. R., A. Fox, and K. F. Fox. 2006. Rapid discrimination of *Bacillus anthracis* from other members of the *B. cereus* group by mass and sequence of "intact" small acid soluble proteins (SASPs) using mass spectrometry. *Journal of Microbiological Methods* 67(2):230–240.

CCD Working Group. 2006. Colony collapse disorder (CCD). http://maarec.cas.psu.edu/ pressReleases/FallDwindleUpdate0107.pdf (accessed April 21, 2009).

Chait, B. T. and S. B. H. Kent. 1992. Weighing naked proteins: Practical, high-accuracy mass measurement of peptides and proteins. *Science* 257(5078):1885–1894.

Chalmers, M. J. and S. J. Gaskel. 2000. Advances in mass spectrometry for proteome analysis. *Current Opinion in Biotechnology* 11(4):384–390.

Chandler, D. P., J. Brown, D. R. Call et al. 2001. Automated immunomagnetic separation and microarray detection of *Escherichia coli* O157:H7 from poultry carcass rinse. *International Journal of Food Microbiology* 70(1–2):143–154.

Chandler, D. P. and A. E. Jarrell. 2004. Automated purification and suspension array detection of 16S rRNA from soil and sediment extracts by using tunable surface microparticles. *Applied and Environmental Microbiology* 70(5):2621–2631.

Chen, R., X. Cheng, D. W. Mitchell et al. 1995. Trapping, detection, and mass determination of Coliphage T4 DNA Ions by electrospray ionization Fourier transform ion cyclotron resonance mass spectrometry. *Analytical Chemistry* 67(7):1159–1163.

Chen, S. 1997. Tandem mass spectrometric approach for determining structure of molecular species of amino phospholipids. *Lipids* 32(1):85–100.

Chen, W., K. E. Laidig, Y. Park et al. 2001. Searching the *Porphyromonas gingivalis* genome with peptide fragmentation mass spectra. *Analyst* 126:52–57.

Chen, Y.P., J. D. Evans, I. B. Smith, and J. S. Pettis. 2008. *Nosema ceranae* is a long-present and wide-spread microsporidian infection of the European honey bee (*Apis mellifera*) in the United States. *Journal of Invertebrate Pathology* 97(2):152–159.

Cheng, J., E. L. Sheldon, L. Wu et al. Preparation and hybridization analysis of DNA/RNA from *Escherichia coli* on microfabricated bioelectronic chips. *Nature Biotechnology* 16:541–546.

Chenna, A. and C. R. Iden. 1993. Characterization of 2′-deoxycytidine and 2′-deoxyuridine adducts formed in reactions with acrolein and 2-bromoacrolein. *Chemical Research in Toxicology* 6(3):261–268.

Chitnis, N. S., S. M. D'Costa, E. R. Paul, and S. L. Bilimoria. 2008. Modulation of iridovirus-induced apoptosis by endocytosis, early expression, JNK, and apical caspase. *Virology* 370(2):333–342.

Chua, J. 2004. A buyer's guide to DNA and RNA prep kits. *Scientist* 18(3):38–39.

Cohen, S. L. and B. T. Chait. 1997. Mass spectrometry of whole proteins eluted from sodium dodecyl sulfate-polyacrylamide gel electrophoresis gels. *Analytical Biochemistry* 247:257–267.

Cole, A. and T. J. Morris. 1980. A new iridovirus of two species of terrestrial isopods, *Armadillidium vulgare* and *Porcellio scaber. Intervirology* 14(1):21–30.

Conway, G. C., S. C. Smole, D. A. Sarracino et al. 2001. Phyloproteomics: Species identification of Enterobacteriaceae using matrix-assisted laser desorption/ionization time-of-flight mass spectrometry. *Journal of Molecular Microbiology and Biotechnology* 3(1):103–112.

Cooks, R. G., Z. Ouyang, Z. Takats, and J. M. Wiseman. 2006. Ambient mass spectrometry. Review. *Science* 311(5767):1566–1570.

Corbin, R. W., O. Paliy, F. Yang et al. 2003. Toward a protein profile of *Escherichia coli*: Comparison to its transcriptome profile. *Proceedings of the National Academy of Sciences of the United States of America* 100(16):9232–9237.

Cotter, R. J. 1992. Time-of-flight mass spectrometry for the structural analysis of biological molecules. *Analytical Chemistry* 64(21):A1027–A1039.

Cox-Foster, D. L., S. Conlan, E. C. Holmes et al. 2007. A metagenomic survey of microbes in honey bee colony collapse disorder. *Science* 318(5848):283–287.

Craft, D. and L. Li. 2005. Integrated sample processing system involving on-column protein adsorption, sample washing, and enzyme digestion for protein identification by LC—ESI MS/MS. *Analytical Chemistry* 77(8):2649–2655.

Craig, R. and R. C. Beavis. 2004. TANDEM: Matching proteins with tandem mass spectra. *Bioinformatics* 20(9):1466–1467.

D'Costa, S. M., H. J. Yao, and S. L. Bilimoria. 2004. Transcriptional mapping in Chilo iridescent virus infections. *Archives of Virology* 149(4):723–742.

Dahouk, S. A., K. Nockler, H. C. Scholz et al. 2006. Immunoproteomic characterization of *Brucella abortus* 1119–3 preparations used for the serodiagnosis of *Brucella* infections. *Journal of Immunological Methods* 309(1–2):34–47.

Dai, J., C. H. Shieh, Q.-H. Sheng, H. Zhou, and R. Zeng. 2005. Proteomic analysis with integrated multiple dimensional liquid chromatography/mass spectrometry based on elution of ion exchange column using pH steps. *Analytical Chemistry* 77(18): 5793–5799.

Daniel, J. M., V. V. Laiko, V. V. Doroshenko, and R. Zenobi. 2005. Interfacing liquid chromatography with atmospheric pressure MALDI-MS. *Analytical and Bioanalytical Chemistry* 383(6):895–902.

Dawson, P. H. 1976. *Quadrupole Mass Spectrometry and Its Application.* Amsterdam, the Netherlands: Elsevier.

De Maesschalck, R., D. Jouan-Rimbaid, and D. L. Massart. 2000. The Mahalanobis distance. *Chemometrics and Intelligent Laboratory Systems* 50(1):1–18.

De Miranda, J. R., M. Drebot, S. Tyler et al. 2004. Complete nucleotide sequence of Kashmir bee virus and comparison with Acute bee paralysis virus. *Journal of General Virology* 85(8):2263–2270.

Debnam, S., D. Westervelt, J. Bromenshenk, and R. Oliver. 2008. Colony collapse disorder: Symptoms change with seasons and are different with various locations. *Bee Culture* 137:30–32.

Delhon, G., E. R. Tulman, C. L. Afonso et al. 2006. Genome of Invertebrate iridescent virus type 3 (Mosquito iridescent virus). *Journal of Virology* 80(17):8439–8449.

DeMarco, D. R. and D. V. Lim. 2002. Detection of *Escherichia coli* O157:H7 in 10- and 25-gram ground beef samples with an evanescent-wave biosensor with silica and polystyrene waveguides. *Journal of Food Protection* 65(4):596–602.

Demirev, P. A., Y.-P. Ho, V. Ryzhov, and C. Fenselau. 1999. Microorganism identification by mass spectrometry and protein database searches. *Analytical Chemistry* 71(14): 2732–2738.

Demirev, P. A., J. S. Lin, F. J. Pineda, and C. Fenselau. 2001a. Bioinformatics and mass spectrometry for microorganism identification: Proteome-wide post-translational modifications and database search algorithms for characterization of intact *Helicobacter pylori*. *Analytical Chemistry* 73(19):4566–4573.

Demirev, P., J. Ramirez, and C. Fenselau. 2001b. Tandem mass spectrometry of intact proteins for characterization of biomarkers from *Bacillus cereus* T spores. *Analytical Chemistry* 73(23):5725–5731.

Deshpande, S.V., R.E. Jabbour, A.P. Snyder et al. 2011. ABOid: A software for automated identification and phyloproteomics classification of tandem mass spectrometric data. *Journal of Chromatography and Separation Techniques* 1(1):S5.

Dickinson, D. N., J. P. Dworzanski, S. V. Deshpande et al. 2005. Classification of BACT group bacteria using an LC-MS/MS based proteomic approach to reveal relatedness between microorganisms. *In Proceedings of the 53rd ASMS Conference on Mass Spectrometry and Allied Topics.* San Antonio, TX, TP27.

Dickinson, D. N., M. T. La Duc, M. Satomi et al. 2004. MALDI-TOF MS compared with other polyphasic taxonomy approaches for the identification and classification of *Bacillus pumilus* spores. *Journal of Microbiological Methods* 58:1–12.

DiGiorgio, C. L., D. A. Gonzalez, and C. C. Huitt. 2002. *Cryptosporidium* and *Giardia* recoveries in natural waters by using Environmental Protection Agency method 1623. *Applied and Environmental Microbiology* 68(12):5952–5955.

Domin, M. A., K. J. Welham, and D. S. Ashton. 1999. The effect of solvent and matrix combinations on the analysis of bacteria by matrix-assisted laser desorption/ionisation time-of-flight mass spectrometry. *Rapid Communications in Mass Spectrometry* 13(4):222–226.

Done, S. H. and I. H. Brown. 1999. Swine influenza viruses in Europe. In *Proceedings of the Allen D. Leman Swine Conference.* 26, 263–267.

Doroshenko, V. M. and R. J. Cotter. 1994. Linear mass calibration in the quadrupole ion-trap mass spectrometer. *Rapid Communications in Mass Spectrometry* 8(9):766–771.

Doroshenko, V. M. and R. J. Cotter. 1996. Advanced stored waveform inverse Fourier transform technique for a matrix-assisted laser desorption/ionization quadrupole ion trap mass spectrometer. *Rapid Communications in Mass Spectrometry* 10(1):65–73.

Duché, O., F. Trémoulet, A. Namane, and J. C. Labadie. 2002. A proteomic analysis of the salt stress response of *Listeria monocytogenes. FEMS Microbiology Letters* 215(2):183–188.

Dumas, M.-E., L. Debrauwer, L. Beyet et al. 2002. Analyzing the physiological signature of anabolic steroids in cattle urine using pyrolysis/metastable atom bombardment mass spectrometry and pattern recognition. *Analytical Chemistry* 74(20):5393–5404.

Dworzanski, J. P., L. Berwald, and H. K. L. Meuzelaar. 1990. Pyrolytic methylation-gas chromatography of whole bacterial cells for rapid profiling of cellular fatty acids. *Applied and Environmental Microbiology* 56(6):1717–1724.

Dworzanski, J. P., S. V. Deshpande, R. Chen et al. 2005. Data mining tools for the classification and identification of bacteria using SEQUEST outputs. In *Proceedings of the 53rd ASMS Conference on Mass Spectrometry and Allied Topics.* San Antonio, TX, TP22.

Dworzanski, J. P., S. V. Deshpande, R. Chen et al. 2006. Mass spectrometry-based proteomics combined with bioinformatics tools for bacterial classification. *Journal of Proteome Research* 5(1):76–87.

Dworzanski, J. P., D. N. Dickinson, S. V. Deshpande et al. 2007. Sequence-based identification and taxonomic classification of microbial agents using MS-based proteomics. Presented at the *5th Annual ASM Biodefense and Emerging Diseases Research Meeting*, Washington, DC, February 27–March 2, 2007.

Dworzanski, J. P., W. H. McClennen, P. A. Cole et al. 1997. Field-portable, automated pyrolysis-GC/IMS system for rapid biomarker detection in aerosols: A feasibility study. *Field Analytical Chemistry and Technology* 1(5):295–305.

Dworzanski, J. P. and H. L. C. Meuzelaar. 1988. Generation of fatty acid methyl ester profiles by direct curie-point pyrolysis GC/MS of whole cell samples. In *Proceedings of the 36th ASMS Conference on Mass Spectrometry and Allied Topics*, June 5–10, 1988. San Francisco, CA: American Society for Mass Spectrometry, pp. 401–402.

Dworzanski, J. P., A. P. Snyder, R. Chen et al. 2004a. Correlation of mass spectrometry identified bacterial biomarkers from a fielded pyrolysis-gas chromatography-ion mobility spectrometry biodetector with the microbiological gram stain classification scheme. *Analytical Chemistry* 76(21):2355–2366.

Dworzanski, J. P., A. P. Snyder, R. Chen et al. 2004b. Identification of bacteria using tandem mass spectrometry combined with a proteome database and statistical scoring. *Analytical Chemistry* 76(8):2355–2366.

Ecelberger, S. A., T. J. Cornish, B. F. Collins et al. 2004. Suitcase TOF: A man-portable time-of-flight mass spectrometer. *Johns Hopkins APL Technical Digest* 25(1):14–19.

Ecker, D. J., J. J. Drader, J. Gutierrez et al. 2006. The Ibis T5000 universal biosensor: An automated platform for pathogen identification and strain typing. *Journal of the Association for Laboratory Automation* 11(6):341–351.

Edgewood Chemical Biological Center. 2007. Scientists Identify Pathogens That May Be Causing Global Honey-Bee Deaths. Science Daily. http://www.sciencedaily.com/releases/2007/04/070426100117.htm.

Emanuel, P. A., J. Dang, J. S. Gebhardt et al. 2000. Recombinant antibodies: A new reagent for biological agent detection. *Biosensors and Bioelectronics* 14(10–11):751–759.

Eng, J. K., A. L. McCormack, and J. R. Yates, III. 1994. An approach to correlate tandem mass spectral data of peptides with amino acid sequences in a protein database. *Journal of the American Society for Mass Spectrometry* 5(11):976–989.

English, R. D., B. Warscheid, C. Fenselau, and R. J. Cotter. 2003. *Bacillus* spore identification via proteolytic peptide mapping with a miniaturized MALDI TOF mass spectrometer. *Analytical Chemistry* 75(24):6886–6893.

Enroth, H. and L. Engstrand. 1995. Immunomagnetic separation and PCR for detection of *Helicobacter pylori* in water and stool specimens. *Journal of Clinical Microbiology* 33(8):2162–2165.

Ethier, F., W. Hou, H. S. Duewel, and D. Figeys. 2006. The proteomic reactor: A microfluidic device for processing minute amounts of protein prior to mass spectrometry analysis. *Journal of Proteome Research* 5(10):2754–2759.

Exner, M. M. and M. A. Lewinski. 2003. Isolation and detection of *Borrelia burgdorferi* DNA from cerebral spinal fluid, synovial fluid, blood, urine, and ticks using the Roche MagNA Pure system and real-time PCR. *Diagnostic Microbiology and Infectious Disease* 46(4):235–240.

Faubert, D., J. G. C. Paul, J. Giroux, and M. J. Bertrand. 1993. Selective fragmentation and ionization of organic compounds using an energy tunable rare-gas metastable beam source. *International Journal of Mass Spectrometry and Ion Physics* 124(1):69–78.

Fenselau, C. and P. A. Demirev. 2001. Characterization of intact microorganisms by MALDI mass spectrometry. *Mass Spectrometry Reviews* 20(4):157–171.

Fergenson, D. P., M. E. Pitesky, H. J. Tobias et al. 2004. Reagentless detection and classification of individual bioaerosol particles in seconds. *Analytical Chemistry* 76(2):373–378.

Fernandez, L. E., H. R. Sorensen, C. Jorgensen et al. 2007. Characterization of oligosaccharides from industrial fermentation residues by matrix-assisted laser desorption/ionization, electrospray mass spectrometry, and gas chromatography mass spectrometry. *Molecular Biotechnology* 35(2):149–160.

Fico, M., M. Yu, Z. Ouyang, R. G. Cooks, and W. J. Chappell. 2007. Miniaturization and geometry optimization of a polymer-based rectilinear ion trap. *Analytical Chemistry* 79(21):8076–8082.

Field, D., B. Tiwari, T. Booth, S. Houten, D. Swan, N. Bertrand, and M. Thurston. 2006. Open software for biologists: From famine to feast. *Nature Biotechnology* 24:801–803.

Fountain, S. T., H. Lee, and D. M. Lubman. 1994. Ion fragmentation activated by matrix-assisted laser desorption/ionization in an ion-trap/reflectron time-of-flight device. *Rapid Communications in Mass Spectrometry* 8(5):487–494.

Fox, A. and G. E. Black. 1994. Identification and detection of carbohydrate markers for bacteria. Derivatization and gas chromatography–mass spectrometry. In *Mass Spectrometry for the Characterization of Microorganisms*, C. Fensealau. ACS Ed. Symposium Series 541. Washington, DC: American Chemical Society. Chapter 8, pp. 107–131.

Fox, A., G. E. Black, K. Fox, and R. Rostovtseva. 1993. Determination of carbohydrate profiles of *Bacillus anthracis* and *Bacillus cereus* including identification of O-methyl methylpentoses by using gas chromatography-mass spectrometry. *Journal of Clinical Microbiology* 31(4):887–894.

Fox, A. and K. Fox. 1991. Rapid elimination of a synthetic adjuvant peptide from the circulation after systemic administration and absence of detectable natural muramyl peptides in normal serum at current analytical limits. *Infectection and Immunity* 59(3):1202–1214.

Fox, A. and S. L. Morgan. 1985. Detection of microorganisms by electrical impedance measurements. In *Instrumental Methods for Rapid Microbiological Analysis*, W. H. Nelson, ed. Deerfield Beach, FA: VCH. Chapter 7, pp. 135–164.

Fraser, C. et al. 2009. Pandemic potential of a strain of Influenza A (H1N1): Early findings. *Science* 324(5934):1557–1561.

Fries, I., F. Feng, A. Da Silva, S. B. Slemenda, and N. J. Pieniazek. 1996. *Nosema ceranae N.* sp. (Microspora, Nosematidae), morphological and molecular characterization of a microsporidian parasite of the Asian honey bee *Apis cerana* (Hymenoptera, Apidae). *European Journal of Protistology* 32(3):356–365.

Fujiyuki, T., S. Ohka, H. Takeuchi et al. 2006. Prevalence and phylogeny of Kakugo virus, a novel insect picorna-like virus that infects the honey bee (*Apis mellifera* L.), under various colony conditions. *Journal of Virology* 80(23):11528–11538.

Gabelica, V., C. Vreuls, P. Filee et al. 2002. Advantages and drawbacks of nanospray for studying noncovalent protein-DNA complexes by mass spectrometry. *Rapid Communications in Mass Spectrometry* 16(18):1723–1728.

Gale, D. C., J. E. Bruce, G. A. Anderson et al. 1993. Bio-affinity characterization mass spectrometry. *Rapid Communications in Mass Spectrometry* 7:1017–1021.

Gantt, S. L., N. B. Valentine, A. J. Saenz et al. 1999. Use of internal control for matrix-assisted laser desorption/ionization time-of-flight mass spectrometry analysis of bacteria. *Journal of the American Society for Mass Spectrometry* 10(11):1131–1137.

Gao, L., Q. Y. Song, G. E. Patterson, R. G. Cooks, and Z. Ouyang. 2006. Handheld rectilinear ion trap mass spectrometer. *Analytical Chemistry* 78(17):5994–6002.

Gärdén, P., R. Alm, and J. Hakkinen. 2005. Proteios: An open source proteomics initiative. *Bioinformatics* 21(9):2085–2087.

Gieray, R. A., P. T. Reilly, M. Yang et al. 1997. Real-time detection of individual airborne bacteria. *Journal of Microbiological Methods* 29(3):191–199.

Goodacre, R., J. K. Heald, and D. B. Kell. 1999. Characterization of intact microorganisms using electrospray ionization mass spectrometry. *FEMS Microbiology Letters* 176(1):17–24.

Gorman, J. J., A. N. Hodder, P. W. Selleck, and E. Hansson. 1992. Antipeptide antibodies for analysis of pathotype-specific variations in cleavage activation of the membrane glycoprotein precursors of Newcastle disease virus isolates in cultured cells. *Journal of Virological Methods* 37(1):55–70.

Griest, W. H. and S. A. Lammert. 2006. The Development of the Block II Chemical Biological Mass Spectrometer. In *Identification of Microorganisms by Mass Spectrometry*, C. L. Wilkins, J. O. Lay, eds. Hoboken, NJ: John Wiley and Sons, Inc., pp. 61–90.

Gu, X., C. Deng, G. Yan, and X. Zhang. 2006. Capillary array reversed-phase liquid chromatography-based multidimensional separation system coupled with maldi-tof-tof-ms detection for high-throughput proteome analysis. *Journal of Proteome Research* 5(11):3186–3196.

Guilhaus, M. 1995. Principles and instrumentation for TOF-MS. *Journal of Mass Spectrometry* 30(11):1519–1532.

Haag, A. M., S. N. Taylor, K. H. Johnston et al. 1998. Rapid identification and speciation of *Haemophilus* bacteria by matrix-assisted laser desorption/ionization time-of-flight mass spectrometry. *Journal of Mass Spectrometry* 33(8):750–756.

Habermann, B., J. Oegema, S. Sunyaev, and A. Shevchenko. 2004. The power and the limitations of cross-species protein identification by mass spectrometry-driven sequence similarity searches. *Molecular and Cellular Proteomics* 3:238–249.

Hager, J. W. and J. C. Le Blanc. 2003. High-performance liquid chromatography-tandem mass spectrometry with a new quadrupole/linear ion trap instrument. *Journal of Chromatography A* 1020(1):3–9.

Halden, R. U., D. R. Colquhoun, and E. S. Wisniewski. 2005. Identification and phenotypic characterization of *Sphingomonas wittichii* strain RW1 by peptide mass fingerprinting using matrix-assisted laser desorption ionization-time of flight mass spectrometry. *Applied and Environmental Microbiology* 71(5):2442–2451.

Harris, W. A. and J. P. Reilly. 2002. On-probe digestion of bacterial proteins for MALDI-MS. *Analytical Chemistry* 74(17):4410–4416.

Hayek, C. S., F. J. Pineda, O. W. Doss, III, and J. S. Lin. 1999. Computer-assisted interpretation of mass spectra. *Johns Hopkins APL Technical Digest* 20(3):363–371.

Henderson, C. W., C. L. Johnson, S. A. Lodhi, and S. L. Bilimoria. 2001. Replication of Chilo iridescent virus in the cotton boll weevil, *Anthonomus grandis*, and development of an infectivity assay. *Archives of Virology* 146(4):767–775.

Hesketh, A. R., G. Chandra, A. D. Shaw et al. 2002. Primary and secondary metabolism, and post-translational protein modifications, as portrayed by proteomic analysis of *Streptomyces coelicolor*. *Molecular Microbiology* 46(4):917–932.

Hettick, J. M., M. L. Kashon, J. P. Simpson et al. 2004. Proteomic profiling of intact Mycobacteria by matrix-assisted laser desorption/ionization time-of-flight mass spectrometry. *Analytical Chemistry* 76(19):5769–5776.

Hillenkamp, F., M. Karas, R. C. Beavis, and B. T. Chait. 1991. Matrix-assisted laser desorption/ionization mass spectrometry of biopolymers. *Analytical Chemistry* 63(24):A1193–A1202.

Hofstadler, S. A., R. Sampath, L. B. Blyn et al. 2005. TIGER: The universal biosensor. *International Journal of Mass Spectrometry* 242(1):23–41.

Holland, R. D., J. G. Wilkes, F. Rafii et al. 1996. Rapid identification of intact whole bacteria based on spectral patterns using matrix-assisted laser desorption/ionization with time-of-flight mass spectrometry. *Rapid Communications in Mass Spectrometry* 10(10):1227–1232.

Höltje, J. V. 1998. Growth of the stress-bearing and shape-maintaining murein sacculus of *Escherichia coli*. *Microbiology and Molecular Biology Reviews* 62(1):181–203.

Honisch, C., Y. Chen, C. Mortimer et al. 2007. Automated comparative sequence analysis by base-specific cleavage and mass spectrometry for nucleic acid-based microbial typing. *Proceedings of the National Academy of Sciences of the United States of America* 104(25):10649–10654.

Hopfgartner, G., C. Husser, and M. Zell. 2003. Rapid screening and characterization of drug metabolites using a new quadrupole-linear ion trap mass spectrometer. *Journal of Mass Spectrometry* 38(2):138–150.

Hopfgartner, G., E. Varesio, V. Tschäppät et al. 2004. Triple quadrupole linear ion trap mass spectrometer for the analysis of small molecules and macromolecules. *Journal of Mass Spectrometry* 39(8):845–855.

Hsu, J., S. J. Chang, and A. H. Franz. 2006. MALDI-TOF and ESI-MS analysis of oligosaccharides labeled with a new multifunctional oligosaccharide tag. *Journal of the American Society for Mass Spectrometry* 17(2):194–204.

Hu, A., P.-J. Tsai, and Y.-P. Ho. 2005. Identification of microbial mixtures by capillary electrophoresis/selective tandem mass spectrometry. *Analytical Chemistry* 77:1488–1495.

Hu, Q., R. Noll, H. Li et al. 2005. The Orbitrap: A new mass spectrometer. *Journal of Mass Spectrometry* 40(4):430–443.

Hua, Y., W. Lu, M. S. Henry et al. 1993. Online high-performance liquid chromatography-electrospray ionization mass spectrometry for the determination of brevetoxins in "Red Tide" algae. *Analytical Chemistry* 67(11):1815–1823.

Hudson, J. R., S. L. Morgan, and A. Fox. 1982. Quantitative pyrolysis gas chromatography-mass spectrometry of bacterial cell walls. *Analytical Biochemistry* 120(1):59–65.

Hukuhara, T. 1964. Induction of epidermal tumor in *Bombyx mori* (Linnaeus) with Tipula iridescent virus. *Journal of Insect Pathology* 6:246–248.

Hunt, D. F. and F. W. Crow. 1978. Electron capture negative ion CI mass spectrometry. *Analytical Chemistry* 50(13):1781–1784.

Hutchens, T. W. and T. T. Yip. 1993. New desorption strategies for the mass spectrometric analysis of macromolecules. *Rapid Communications in Mass Spectrometry* 7(7): 576–580.

Ibekwe, A. M. and C. M. Grieve. 2003. Detection and quantification of *Escherichia coli* O157:H7 in environmental samples by real-time PCR. *Journal of Applied Microbiology* 94(3):421–431.

Ihling, C. and A. Sinz. 2005. Proteome analysis of *Escherichia coli* using high-performance liquid chromatography and Fourier transform ion cyclotron resonance mass spectrometry. *Proteomics* 5:2029–2042.

Inglis, T. J. J., M. Aravena-Roman, S. Ching et al. 2003. Cellular fatty acid profile distinguishes *Burkholderia pseudomallei* from avirulent *Burkholderia thailandensis*. *Journal of Clinical Microbiology* 41(10):4812–4814.

Ivnitski, D., D. J. O'Neil, A. Gattuso et al. 2003. Nucleic acid approaches for detection and identification of biological warfare and infectious disease agents. *BioTechniques* 35:862–869.

Jabbour, R. E., S. V. Deshpande, M. F. Stanford, C. H. Wick, A. W. Zulich, and A. P. Snyder. 2011. A protein processing filter method for bacterial identification by mass spectrometry-based proteomics. *Journal of Proteome Research* 10:907–912.

Jabbour, R. E., S. V. Deshpande, M. M. Wade et al. 2010b. Double-blind characterization of non-genome-sequenced bacteria by mass spectrometry-based proteomics. *Applied and Environmental Microbiology* 76(11):3637–3644.

Jabbour, R., J. P. Dworzanski, S. V. Deshpande et al. 2005. Effect of gas phase fractionation of peptide ions on bacterial identification using mass spectrometry-based proteomics approach. In the *Proceedings of the 53rd ASMS Conference on Mass Spectrometry and Allied Topics*, San Antonio, TX, TP31.

Jabbour, R. E., J. P. Dworzanski, S. V. Deshpande et al. 2007. Effect of microbial sample processing conditions on bacterial identification using mass spectrometry-based proteomics approach. In *Proceedings of the 55th ASMS Conference in Mass Spectrometry and Allied Topics*, June 2007, Indianapolis, Indiana.

Jabbour, R. E., J. P. Dworzanski, S. V. Deshpande et al. 2008. Processing of microbial samples using protein ultrafiltration devices for identification of bacteria by mass spectrometry. In *Proceedings of the 56th ASMS Conference in Mass Spectrometry and Allied Topics*, June 2008, Denver, CO.

Jabbour, R. E., M. M. Wade, S. V. Deshpande et al. 2010a. Identification of *Yersinia pestis* and *Escherichia coli* strains by whole cell and outer membrane protein extracts with mass spectrometry-based proteomics. *Journal of Proteome Research* 9(7):3647–3655.

Jackson, G. W., R. J. McNichols, G. E. Fox, and R. C. Willson. 2007. Universal bacterial identification by mass spectrometry of 16S ribosomal RNA cleavage products. *International Journal of Mass Spectrometry* 261:218–226.

Jaffe, J. D., H. C. Berg, and G. M. Church. 2004. Proteogenomic mapping as a complementary method to perform genome annotation. *Proteomics* 4(1):59–77.

Jaffer, A. S. 2004. Biohazard detection system deployment resumes. U. S. Postal Service Postal News 38.

Jakob, N. J., K. Müller, U. Bahr, and G. Darai. 2001. Analysis of the first complete DNA sequence of an invertebrate iridovirus: Coding strategy of the genome of Chilo iridescent virus. *Virology* 286(1):182–196.

Janini, G. M., T. P. Conrads, K. L. Wilkens et al. 2003. A sheathless nanoflow electrospray interface for on-line capillary electrophoresis mass spectrometry. *Analytical Chemistry* 75(7):1615–1619.

Janini, G. M., M. Zhou, L.-R. Yu et al. 2003. On-column sample enrichment for capillary electrophoresis sheathless electrospray ionization mass spectrometry: Evaluation for peptide analysis and protein identification. *Analytical Chemistry* 75(21):5984–5993.

Jantzen, E. 1984. Application of GC-MS to the study of amino acids and peptides. In *Gas Chromatography/Mass Spectrometry Applications in Microbiology*, G. Odham, L. Larson, and P. A. Mardh, eds. New York: Plenum Press. Chapter 8, pp. 257–302.

Jarman, K. H., S. T. Cebula, A. J. Saenz et al. 2000. An algorithm for automated bacterial identification using matrix-assisted laser desorption/ionization time-of-flight mass spectrometry. *Analytical Chemistry* 72(6):1217–1223.

Jarman, K. H., D. S. Daly, C. E. Petersen et al. 1999. Extracting and visualizing matrix-assisted laser desorption/ionization time-of-flight mass spectral fingerprints. *Rapid Communications in Mass Spectrometry* 13(15):1586–1594.

Jones, J. J., M. J. Stump, R. C. Fleming et al. 2003. Investigation of MALDI-TOF and FT-MS techniques for analysis of *Escherichia coli* whole cells. *Analytical Chemistry* 75(6):1340–1347.

Jonscher, K. R. and J. R. Yates, III. 1996. Mixture analysis using a quadrupole mass filter/quadrupole ion trap mass spectrometer. *Analytical Chemistry* 68(4):659–667.

Karas, M. and F. Hillenkamp. 1988. Laser desorption ionization of proteins with molecular masses exceeding 10,000 daltons. *Analytical Chemistry* 60(20):259–280.

Karataev, V. I., B. A. Mamyrin, and D. V. Shmikk. 1972. New method for focusing ion bunches in time-of-flight mass spectrometers. *Soviet Physics–Technical Physics* 16:1177–1179.

Keil, A., N. Talaty, C. Janfelt et al. 2007. Ambient mass spectrometry with a handheld mass spectrometer at high pressure. *Analytical Chemistry* 79(20):7734–7739.

Kelleher, N. L. 2004. Top-down proteomics. *Analytical Chemistry* 76(11):196A–203A.

Keller, A., A. I. Nesvizhskii, E. Kolker, and R. Aebersold. 2002. Empirical statistical model to estimate the accuracy of peptide identifications made by MS/MS and database search. *Analytical Chemistry* 74(20):5383–5392.

Khalsa-Moyers, G. and W. H. McDonald. 2006. Developments in mass spectrometry for the analysis of complex protein mixtures. *Briefings in Functional Genomics and Proteomics* 5:98–111.

Kievit, O., M. Weiss, P. J. T. Verheijen, J. C. M. Marijnissen, and B. Scarlett. 1996. The online chemical analysis of single particles using aerosol beams and time of flight mass spectroscopy. *Chemical Engineering Communications* 151(1):79–100.

King, R. and C. Fernandez-Metzler. 2006. The use of QTrap technology in drug metabolism. *Current Drug Metabolism* 7(5):541–545.

Kirby, R., E. J. Cho, B. Gehrkem et al. 2004. Aptamer-based sensor arrays for the detection and quantitation of proteins. *Analytical Chemistry* 76(14):4066–4075.

Klietmann, W. F. and K. L. Ruoff. 2001. Bioterrorism: Implications for the clinical microbiologist. *Clinical Microbiology Reviews.* 14(2):364–381.

Ko, M., H. Choi, and C. Park. 2002. Group I self-splicing intron in the recA gene of *Bacillus anthracis. Journal of Bacteriology* 184(14):3917–3922.

Kolker, E., A. F. Picone, M. Y. Galperin et al. 2005. Global profiling of *Shewanella oneidensis* MR-1: Expression of hypothetical genes and improved functional annotations. *Proceedings of the National Academy of Sciences of the United States of America* 102(6):2099–2104.

Kolker, E., S. Purvine, M. Y. Galperin et al. 2003. Initial proteome analysis of model microorganism *Haemophilus influenzae* strain Rd KW20. *Journal of Bacteriology* 185(15):4593–4602.

Krishnamurthy, T., P. L. Ross, and U. Rajamani. 1996. Detection of pathogenic and nonpathogenic bacteria by matrix-assisted laser desorption/ionization time-of-flight mass spectrometry. *Rapid Communications in Mass Spectrometry* 10(8):883–888.

Kwon, Y., K. Tang, C. Cantor, H. Koster, and C. Kang. 2002. DNA sequencing and genotyping by transcriptional synthesis of chain-terminated RNA ladders and MALDI-TOF mass spectrometry. *Nucleic Acids Research* 29(3):E11.

La Duc, M. T., M. Satomi, N. Agata, and K. Venkateswaran. 2004. gyrB as a phylogenetic discriminator for members of the *Bacillus anthracis-cereus-thuringiensis* group. *Journal of Microbiological Methods* 56(3):383–394.

Laiko, V. V., S. C. Moyer, and R. J. Cotter. 2000. Atmospheric pressure MALDI/ion trap mass spectrometry. *Analytical Chemistry* 72(21):5239–5243.

Lambert, J.-P., M. Ethier, J. C. Smith, D. Figeys. 2005. Proteomics: From gel based to gel free. *Analytical Chemistry* 77(12):3771–3788.

Lampel, K., D. Dyer, L. Kornegay, L., and P. Orlandi. 2004. Detection of Bacillus spores using PCR and FTA filters. *Journal of Food Protrection* 67(5):1036–1038.

Lancashire, L., O. Schmid, H. Shah, and G. Ball. 2005. Classification of bacterial species from proteomic data using combinatorial approaches incorporating artificial neural networks, cluster analysis and principal component analysis. *Bioinformatics* 21(10): 2191–2199.

Lay, J. O., Jr. 2001. MALDI-TOF mass spectrometry of bacteria. *Mass Spectrometry Reviews* 20(4):172–194.

Leclercq, A., A. Guiyoule, M. El Lioui, E. Carniel, and J. Decallonne. 2000. High homogeneity of the *Yersinia pestis* fatty acid composition. *Journal of Clinical Microbiology* 38(4):1545–1551.

Lee, H. and S. K. R. Williams. 2003. Analysis of whole bacterial cells by flow field-flow fractionation and matrix-assisted laser desorption/ionization time-of-flight mass spectrometry. *Analytical Chemistry* 75(11):2746–2752.

Lee, J.-G., K. H. Cheong, N. Huh, S. Kim, J.-W. Choi, and C. Ko. 2006. Microchip-based one step DNA extraction and real-time PCR in one chamber for rapid pathogen identification. *Lab on a Chip* 6(7):886–895.

Lee, K., D. Bae, and D. Lim. 2002. Evaluation of parameters in peptide mass fingerprinting for protein identification by MALDI-TOF mass spectrometry. *Molecules and Cells* 13(2):175–184.

Leenders, F., T. H. Stein, B. Kablitz, P. Franke, and J. Vater. 1999. Rapid typing of *Bacillus subtilis* strains by their secondary metabolites using matrix-assisted laser desorption/ionization mass spectrometry of intact cells. *Rapid Communications in Mass Spectrometry* 13(10):943–949.

Lefmann, M., C. Honisch, S. Bocker et al. 2004. Novel mass spectrometry-based tool for genotypic identification of mycobacteria. *Journal of Clinical Microbiology* 42(1):339–346.

Li, J., Z. Zhang, J. Rosenzweig et al. 2002. Proteomics and bioinformatics approaches for identification of serum biomarkers to detect breast cancer. *Clinical Chemistry* 48(8):1296–1304.

Li, Y., F. Wenzel, W. Holzgreve, and S. Hahn. 2006. Genotyping fetal paternally inherited SNPs by MALDI-TOF MS using cell-free fetal DNA in maternal plasma: Influence of size fractionation. *Electrophoresis* 27(19):3889–3896.

Licklider, L., X. Wang, A. Desai, Y. Tai, and T. Lee. 2000. A micromachined chip-based electrospray source for mass spectrometry. *Analytical Chemistry* 72(2):367–375.

Lin, S. S., C. H. Wu, M. C. Sun, C. M. Sun, and Y.-P. Ho. 2005. Microwave-assisted enzyme-catalyzed reactions in various solvent system. *Journal of the American Society for Mass Spectrometry* 16(4):581–588.

Lin, W. T., W. N. Hung, Y. H. Yian et al. 2006. Multi-Q: A fully automated tool for multiplexed protein quantitation. *Journal of Proteome Research* 5(9):2328–2338.

Lin, Y.-S., P.-J. Tsai, M.-F. Weng, and Y.-C. Chen. 2005. Affinity capture using vancomycin-bound magnetic nanoparticles for the maldi-ms analysis of bacteria. *Analytical Chemistry* 77(6):1753–1760.

Linde, H.-J., H. Neubauer, H. Meyer et al. 1999. Identification of *Yersinia* species by the Vitek GNI card. *Journal of Clinical Microbiology* 37(1):211–214.

Lipton, M. S., L. Pasa-Tolic, G. A. Anderson et al. 2002. Global analysis of the *Deinococcus radiodurans* proteome by using accurate mass tags. *Proceedings of National Academy of Sciences USA* 99(17):11049–11054.

Liska, A. J. and A. Shevchenko. 2003. Combining mass spectrometry with database interrogation strategies in proteomics. *Trends in Analytical Chemistry* 22(5):291–298.

Liu, H., Z. Du, J. Wang, and R. Yang. 2007. Universal sample preparation method for characterization of bacteria by matrix-assisted laser desorption ionization–time of flight mass spectrometry. *Applied and Environmental Microbiology* 73(6):1899–1907.

Liu, J., K. W. Ro, M. Busman, and D. R. Knapp. 2004. Electrospray ionization with a pointed carbon fiber emitter. *Analytical Chemistry* 76(13):3599–3606.

Liu, T.-Y., L.-L. Shiu, T.-Y. Luh, and G.-R. Her. 1995. Direct analysis of C60 and related compounds with electrospray mass spectrometry. *Rapid Communications in Mass Spectrometry* 9(1):93–96.

Lopez, M. F. 2000. Better approaches to finding the needle in a haystack: Optimizing proteome analysis through automation. *Electrophoresis* 21:1082–1093.

Lopez, M., J. C. Rojas, R. Vandame, and T. Williams. 2002. Parasitoid-mediated transmission of an iridescent virus. *Journal of Invertebrate Pathology* 80(3):160–170.

Lou, Q., K. Tang, F. Yang et al. 2006. More sensitive and quantitative proteomic measurements using very low flow rate porous silica monolithic LC columns with electrospray ionization-mass spectrometry. *Journal of Proteome Research* 5(5):1091–1097.

Ludwig, W., O. Strunk, S. Klugbauer et al. 1998. Bacterial phylogeny based on comparative sequence analysis. *Electrophoresis* 19(4):554–568.

Luna, V. A., D. C. King, T. Rycerz et al. 2003. Novel sample preparation method for safe and rapid detection of *Bacillus anthracis* spores in environmental powders and nasal swabs. *Journal of Clinical Microbiology* 41(3):1252–1255.

Lundquist, M., M. B. Caspersen, P. Wikstrom, and M. Forsman. 2005. Discrimination of *Francisella tularensis* subspecies using surface enhanced laser desorption ionization mass spectrometry and multivariate data analysis. *FEMS Microbiology Letters* 243(1):303–310.

Ma, J., J. Liu, L. Sun et al. 2009. Online integration of multiple sample pretreatment steps involving denaturation, reduction, and digestion with microflow reversed-phase liquid chromatography—electrospray ionization tandem mass spectrometry for high-throughput proteome profiling. *Analytical Chemistry* 81(15):6534–6540.

Macek, B., L. F. Waanders, J. V. Olsen, and M. Mann. 2006. Top-down protein sequencing and MS3 on a hybrid linear quadrupole ion trap-orbitrap mass spectrometer. *Molecular and Cellular Proteomics* 5(5):949–958.

Madonna, A. J., F. Basile, and K. J. Voorhees. 2001. Detection of bacteria from biological mixtures using immunomagnetic separation combined with matrix-assisted laser desorption/ionization time-of-flight mass spectrometry. *Rapid Communications in Mass Spectrometry* 15(13):1068–1074.

Madonna, A. J., S. Van Cuyk, and K. J. Voorhees. 2003. Detection of *Escherichia coli* using immunomagnetic separation and bacteriophage amplification coupled with matrix-assisted laser desorption/ionization time-of-flight mass spectrometry. *Rapid Communications in Mass Spectrometry* 17(3):257–263.

Makarov, A., E. Denisov, A. Kholomeev et al. 2006b. Performance evaluation of a hybrid linear ion trap/orbitrap mass spectrometer. *Analytical Chemistry* 78(7):2113–2120.

Makarov, A., E. Denisov, O. Lange, and S. Horning. 2006a. Dynamic range of mass accuracy in LTQ Orbitrap hybrid mass spectrometer. *Journal of the American Society for Mass Spectrometry* 17(7):977–982.

Malen, H., F. S. Berven, T. Softeland et al. 2008. Membrane and membrane-associated proteins in Triton X-114 extracts of *Mycobacterium bovis* BCG identified using a combination of gel-based and gel-free fractionation strategies. *Proteomics* 8:1859–1870.

Mamyrin, B. A. 2001. Time-of-flight mass spectrometry (concepts, achievements and prospects). *International Journal of Mass Spectrometry* 206(3):251–266.

Mamyrin, B. A., V. I. Karataev, D. V. Shmikk, and V. A. Zagulin. 1973. Reflectron for TOF-MS. *Soviet Physics Journal of Experimental and Theoretical Physics* 37:45–48.

Mandrell, R. E., L. A. Harden, A. Bates et al. 2005. Speciation of *Campylobacter coli, C. jejuni, C. helveticus, C. lari, C. sputorum, and C. upsaliensis* by Matrix-Assisted Laser Desorption Ionization-Time of Flight Mass Spectrometry. *Applied and Environmental Microbiology* 71(10):6292–6307.

Mann, M. and M. S. Wilm, 1994. Error-tolerant identification of peptides in sequence databases by peptide sequence tags. *Analytical Chemistry* 66(24):4390–4399.

Maori, E., S. Lavi, R. Mozes-Koch et al. 2007. Isolation and characterization of Israeli acute paralysis virus, a dicistrovirus affecting honey bees in Israel: Evidence for diversity due to intra- and inter-species recombination. *Journal of General Virology* 88(12):3428–3438.

Maori, E., N. Paldi, S. Shafir et al. 2009. IAPV, a bee-affecting virus associated with colony collapse disorder can be silenced by dsRNA ingestion. *Insect Molecular Biology* 18(1):55–60.

Marijnissen, J., B. Scarlett, and P. Verheijen. 1988. Proposed on-line aerosol analysis combining size determination, laser-induced fragmentation and time-of-flight mass spectroscopy. *Journal of Aerosol Science* 19(7):1307–1310.

Marina, C. F., J. Arredondo-Jimenez, A. Castillo, and T. Williams. 1999. Sublethal effects of iridovirus disease in a mosquito. *Oecologia* 119(3):383–388.

Marina, C. F., I. Fernández-Salas, J. E. Ibarra et al. 2005. Transmission dynamics of an iridescent virus in an experimental mosquito population: The role of host density. *Ecological Entomology* 30(4):376–382.

Marina, C. F., J. E. Ibarra, J. I. Arredondo-Jimenez et al. 2003. Adverse effects of covert iridovirus infection on the life history and demographic parameters of *Aedes aegypti*. *Entomologia Experimentalis et Applicata* 106(1):53–61.

Martin, S. E., J. Shabanowitz, D. F. Hunt, and J. Marto. 2000. Subfemtomole MS and MS/MS peptide sequence analysis using nano-HPLC micro-ESI Fourier transform ion cyclotron resonance mass spectrometry. *Analytical Chemistry* 72(18):4266–4274.

Mason, H.-Y., C. Lloyd, M. Dice, R. Sinclair, W. Ellis, Jr., and L. Powers. 2003. Taxonomic identification of microorganisms by capture and intrinsic fluorescence detection. *Biosensor and Bioelectronics* 18(5–6):521–527.

Mayr, B., U. Kobold, M. Moczko, A. Nyeki, T. Koch, and C. Huber. 2005. Identification of bacteria by polymerase chain reaction followed by liquid chromatography-mass spectrometry. *Analytical Chemistry* 77(14):4563–4570.

McBride, M. T., D. Masquelier, B. J. Hindson et al. 2003. Autonomous detection of aerosolized *Bacillus anthracis* and *Yersinia pestis*. *Analytical Chemistry* 75(20):5293–5299.

McCluckey, S. A., G. Vaidyanathan, and S. Habibi-Goudarzi. 1995. Charged vs. neutral nucleobase loss from multiply charged oligonucleotide anions. *Journal of Mass Spectrometry* 30(9):1222–1229.

McLean, J. A., B. T. Ruotolo, K. J. Gillig, and D. H. Russell. 2005. Ion mobility-mass spectrometry: A new paradigm for proteomics. *International Journal of Mass Spectrometry* 240(3):301–315.

McLoughlin, M. P., W. R. Allmon, C. W. Anderson, M. A. Carlson, D. J. DeCicco, and N. H. Evancich. 1999. Development of a field-portable time-of-flight mass spectrometer system. *Johns Hopkins APL Technical Digest* 20(3):326–334.

McMillan, W. 2002. Real-time point-of-care molecular detection of infectious disease agents. *American Clinical Laboratory* 21:29–31.

Meetani, M. A., Y.-S. Shin, S. Zhang et al. 2007. Desorption electrospray ionization mass spectrometry of intact bacteria. *Journal of Mass Spectrometry* 42(9):1186–1193.

Meuzelaar, H. L. C., J. Haverkamp, and F. D. Hileman. 1982. *Pyrolysis Mass Spectrometry of Recent and Fossil Biomaterials; Compendium and Atlas*. Amsterdam, the Netherlands: Elsevier Science.

Michel, J. 2008. Scientists discover new virus invading US honeybees. U.S. Army Edgewood Chemical Biological Center, Public Affairs Office. News Release. http://www.ecbc. army.mil/pr/download/VDV-1_ Discvery.pdf.

Moe, M. K., T. Anderssen, M. B. Strom, and E. Jensen. 2005. Total structure characterization of unsaturated acidic phospholipids provided by vicinal di-hydroxylation of fatty acid double bonds and negative electrospray ionization mass spectrometry. *Journal of the American Society for Mass Spectrometry* 16(1):46–59.

Morris, H. R., T. Paxton, A. Dell et al. 1996. High sensitivity collisionally-activated decomposition tandem mass spectrometry on a novel quadrupole/orthogonal-acceleration time-of-flight mass spectrometer. *Rapid Communications in Mass Spectrometry* 10(8): 889–896.

Murray, P. R., E. J. Baron, J. H. Jorgensen, M. A. Pfaller, and R. H. Yolken, eds. 2003. *Manual of Clinical Microbiology*. 8th edn. Washington, DC: ASM Press.

National Center for Biotechnology Information. 2009. Influenza virus resource: Information, search and analysis. http://www.ncbi.nlm.nih.gov/genomes/FLU/ SwineFlu.html (accessed May 12, 2009).

National Center for Biotechnology Information. 2010. Influenza virus resource: Information, search and analysis. http://www.ncbi.nlm.nih.gov/genomes/FLU (accessed March 16, 2009).

Nesvizhskii A. I., A. Keller, E. Kolker, and R. Aebersold. 2003. A statistical model for identifying proteins by tandem mass spectrometry. *Analytical Chemistry* 75(17): 4646–4658.

Norbeck, A. D., S. J. Callister, M. E. Monroe et al. 2006. Proteomic approaches to bacterial differentiation. *Journal of Microbiological Methods* 67(3):473–486.

Novel Swine-Origin Influenza A (H1N1) Virus Investigation Team, F. S. Dawood, S. Jain, L. Finelli, M. W. Shaw, S. Lindstrom, R. J. Garten, L. V. Gubareva, X. Xu, C. B. Bridges, T. M. Uyeki. 2009. Emergence of a novel swine-origin influenza A (H1N1) virus in humans. *The New England Journal of Medicine* 360(25):2605–2615.

Odumeru, J. A., M. Steele, L. Fruhner et al. 1999. Evaluation of accuracy and repeatability of identification of food-borne pathogens by automated bacterial identification systems. *Journal of Clinical Microbiology* 37(4):944–949.

Oktem, B., A. K. Sundaram, S. Shanbhag, C. M. Murphy, and V. M. Doroshenko. 2007. High throughput sample preparation for atmospheric pressure MALDI-MS for rapid detection and identification of microorganisms. Presented at the *55th Conference of the American Society for Mass Spectrometry*, 3–7 June, 2007, Indianapolis, IN, Abstract ThP 157.

Ongus, J. R., E. C. Roode, C. W. A. Pleij, J. M. Vlak, and M. M. van Oers. 2006. The 59 non-translated region of Varroa destructor virus 1 (genus *Iflavirus*): Structure prediction and IRES activity in Lymantria dispar cells. *Journal of General Virology* 87:3397–3407.

Owen, R. J., M. A. Claydon, J. Gibson et al. 1999. Strain variation within *Helicobacter pylori* detected by mass spectrometry of cell wall surfaces. *Gut* 45(3):A28.

Owens, D. R., B. Bothner, Q. Phung, K. Harris, and G. Siuzdak. 1998. Aspects of oligo-nucleotide and peptide sequencing with MALDI and electrospray mass spectrometry. *Bioorganic and Medicinal Chemistry* 6(9):1547–1554.

Patterson, G. E., A. J. Guymon, L. S. Riter et al. 2002. A miniature cylindrical ion trap mass spectrometer. *Analytical Chemistry* 74(24):6145–6153.

Paul, E. R., N. S. Chitnis, C. W. Henderson et al. 2007. Induction of apoptosis by iridovirus protein extract. *Archives of Virology* 152(7):1353–1364.

Perch-Nielsen, I. R., D. D. Bang, C. R. Poulsen, J. El-Ali, and A. Wolff. 2003. Removal of PCR inhibitors using dielectrophoresis as a selective filter in a microsystem. *Lab on a Chip* 3:212–216.

Perkins, D. N., D. J. Pappin, D. M. Creasy, and J. S. Cottrell. 1999. Probability-based protein identification by searching sequence databases using mass spectrometry data. *Electrophoresis* 20(18):3551–3567.

Peterson, D. S., T. Rohr, F. Svec, and J. M. J. Frechet. 2003. Dual-function microanalytical device by in situ photolithographic grafting of porous polymer monolith: Integrating solid-phase extraction and enzymatic digestion for peptide mass mapping. *Analytical Chemistry* 75(20):5328–5335.

Petrenko, V. A. and I. B. Sorokulova. 2004. Detection of biological threats: A challenge for directed molecular evolution. *Journal of Microbiological Methods* 58(2):147–168.

Phan, H. N. and A. R. McFarland. 2004. Aerosol-to-hydrosol transfer stages for use in bioaerosol sampling. *Aerosol Science and Technology* 38(4):300–310.

Pierce, C. Y., J. R. Barr, A. R. Woolfitt et al. 2007. Strain and phase identification of the U.S. category B agent *Coxiella burnetii* by matrix assisted laser desorption/ionization time-of-flight mass spectrometry and multivariate pattern recognition. *Analytica Chimica Acta* 583(1):23–31.

Pietsch, C., C. Wiegand, M. V. Ame, A. Nicklisch, D. Wunderlin, and S. Pflugmacher. 2001. The effects of a cyanobacterial crude extract on different aquatic organisms: Evidence for cyanobacterial toxin modulating factors. *Environmental Toxicology* 16(6):535–542.

Pineda, F. J., M. D. Antoine, P. A. Demirev et al. 2003. Microorganism identification by matrix-assisted laser/desorption ionization mass spectrometry and model-derived ribosomal protein biomarkers. *Analytical Chemistry* 75(15):3817–3822.

Pineda, F. J., J. S. Lin, C. Fenselau, and P. A. Demirev. 2000. Testing the significance of microorganism identification by mass spectrometry and proteome database search. *Analytical Chemistry* 72(16):3739–3744.

Pinkston, J. D., T. E. Delaney, D. J. Bowling, and T. L. Chester. 1991. Comparison by capillary SFC and SFC-MS of soxhlet and supercritical fluid extraction of hamster feces. *High Resolution Chromatography* 14(6):401–406.

Poon, T. C. W., K. C. Chan, and P. C. Ng. 2004. Serial analysis of plasma proteomic signatures in pediatric patients with severe acute respiratory syndrome and correlation with viral load. *Clinical Chemistry* 50(8):1452–1455.

Pribil, P. A., E. Patton, G. Black, V. Doroshenko, and C. Fenselau. 2005. Rapid characterization of *Bacillus* spores targeting species-unique peptides produced with an atmospheric pressure matrix-assisted laser desorption/ionization source. *Journal of Mass Spectrometry* 40(4):464–474.

Qian, M. G. and D. M. Lubman. 1995. Analysis of tryptic digests using microbore HPLC with an ion trap storage/reflectron time-of-flight detector. *Analytical Chemistry* 67(17): 2870–2877.

Quadroni, M. and P. James. 1999. Proteomics and automation. *Electrophoresis* 20(4–5): 664–677.

Rådström, P., R. Knutsson, P. Wolffs, M. Lovenklev, and C. Lofstrom. 2004. Pre-PCR processing: Strategies to generate PCR-compatible samples. *Molecular Biotechnology* 26(2):133–146.

Ramsey, R. S. and J. M. Ramsey. 1997. Generating electrospray from microchip devices using electroosmotic pumping. *Analytical Chemistry* 69(6):1174–1178.

Rauch, A., M. Bellew, J. Eng et al. 2006. Computational proteomics analysis system (CPAS): An extensible, open-source analytic system for evaluating and publishing proteomic data and high throughput biological experiments. *Journal of Proteome Research* 5(1):112–121.

Real-Time, R. T. P. C. R. 2009. Emergence of a novel swine-origin influenza A (H1N1) virus in humans. *New England Journal of Medicine* 360:2605–2615.

Reid, A. H., T. G. Fanning, J. V. Hultin, and J. K. Taubenberger. 1999. Origin and evolution of the 1918 "Spanish" influenza virus hemagglutinin gene. *Proceedings of the National Academy of Sciences* 96(4):1651–1656.

Reschiglian, P., A. Zattoni, L. Cinque et al. 2004. Hollow-fiber flow field-flow fractionation for whole bacteria analysis by matrix-assisted laser desorption/ionization time-of-flight mass spectrometry. *Analytical Chemistry* 76(7):2103–2111.

Romay, F. J., D. L. Roberts, V. A. Marple, B. Y. H. Liu, and B. A. Olson. 2002. A high-performance aerosol concentrator for biological agent detection. *Aerosol Science Technology* 36(2):217–226.

Ruelle, V., B. E. Moualij, W. Zorzi, P. Ledent, and E. De Pauw. 2004. Rapid identification of environmental bacterial strains by matrix-assisted laser desorption/ionization time-of-flight mass spectrometry. *Rapid Communications in Mass Spectrometry* 18(18):2013–2019.

Russell, S. C., G. Czerwieniec, C. Lebrilla et al. 2005. Achieving high detection sensitivity (14 zmol) of biomolecular ions in bioaerosol mass spectrometry. *Analytical Chemistry* 77(15):4734–4741.

Russell, S. C., G. Czerwieniec, H. Tobias et al. 2004. Toward understanding the ionization of biomarkers from micrometer particles by bio-aerosol mass spectrometry. *Journal of the American Society for Mass Spectrometry* 15(6):900–909.

Ryzhov, V., Y. Hathout, and C. Fenselau. 2000. Rapid characterization of spores of *Bacillus cereus* group bacteria by matrix-assisted laser desorption-ionization time-of-flight mass spectrometry. *Applied and Environmental Microbiology* 66(9):3828–3834.

Sadygov, R. G., D. Cociorva, and J. R. Yates, III. 2004. Large-scale database searching using tandem mass spectra: Looking up the answer in the back of the book. *National Methods* 1:195–202.

Saikaly, P. E., M. A. Barlaz, and F. L. de los Reyes, III. 2007. Development of quantitative real-time PCR assays for detection and quantification of surrogate biological warfare agents in building debris and leachate. *Applied and Environmental Microbiology* 73(20):6557–6565.

Salzano, A. M., S. Arena, G. Renzone et al. 2007. A widespread picture of the *Streptococcus thermophilus* proteome by cell lysate fractionation and gel-based/gel-free approaches. *Proteomics* 7:1420–1433.

Sampath, R., S. A. Hofstadler, L. B. Blyn et al. 2005. Rapid identification of emerging pathogens: Coronavirus. *Emerging Infectious Diseases* 11(3):373–379.

Sampath, R., K. L. Russell, C. Massire et al. 2007. Global surveillance of emerging influenza virus genotypes by mass spectrometry. *PLoS ONE* 5:e489.

Scaramozzino, N., J.-M. Crance, A. Jouan, D. A. DeBriel, F. Stoll, and D. Garin. 2001. Comparison of Flavivirus universal primer pairs and development of a rapid, highly sensitive heminested reverse transcription-PCR assay for detection of Flaviviruses targeted to a conserved region of the NS5 gene sequences. *Journal of Clinical Microbiology* 39(5):1922–1927.

Schmidt, A., M. Karas, and T. Dülcks. 2003. Effect of different solution flow rates on analyte ion signals in nano-ESI MS, or: When does ESI turn into nano-ESI? *Journal of the American Society for Mass Spectrometry* 14(5):492–500.

Schneider, B. B., V. I. Baranov, H. Javaheri, and T. R. Covey. 2003. Particle discriminator interface for nanoflow ESI-MS. *Journal of the American Society for Mass Spectrometry* 14(11):1236–1246.

Schnelle, T., T. Muller, G. Gradl, S. G. Shirley, and G. Fuhr. 2000. Dielectrophoretic manipulation of suspended submicron particles. *Electrophoresis* 21(1):66–73.

Scholtissek, C., V. S. Hinshaw, and C. W. Olsen. 1998. Influenza in pigs and their role as the intermediate host. In *Textbook of Influenza* K. G. Nicholson, R. G. Webster, and A. J. Hay, eds. Oxford, U.K.: Blackwell Science, 137–145.

Schultz, U., W. M. Fitch, S. Ludwig, J. Mandler, and C. Scholtissek. 1991. Evolution of pig influenza viruses. *Virology* 183(1):61–73.

Shevchenko, A., S. Sunyaev, A. Liska, P. Bork, and A. Shevchenko. 2002. Nanoelectrospray tandem mass spectrometry and sequence similarity searching for identification of proteins from organisms with unknown genomes. *Methods in Molecular Biology* 211:221–234.

Teramoto, K., W. Kitagawa, H. Sato et al. 2009. Phylogenetic analysis of *Rhodococcus eryth-ropolis* based on the variation of ribosomal proteins as observed by matrix-assisted laser desorption ionization-mass spectrometry without using genome information. *Journal of Bioscience and Bioengineering* 108(4):348–353.

Teramoto, K., H. Sato, L. Sun et al. 2007. Phylogenetic classification of *Pseudomonas putida* strains by MALDI-MS using ribosomal subunit proteins as biomarkers. *Analytical Chemistry* 79(22):8712–8719.

Terio, V., V. Martella, M. Camero et al. 2008. Detection of a honey bee iflavirus with intermediate characteristics between KaKugo virus and deformed wing virus. *New Microbiology* 31(4):439–444.

Timmins, E. and R. Goodacre. 2006. Discrimination and identification of microorganisms by pyrolysis mass spectrometry: From burning ambitions to cooling embers–historical perspective. In *Identification of Microorganisms by Mass Spectrometry*, C. L. Wilkins, and J. O. Lay, Jr., eds, New York: John Wiley & Sons, Inc. Chapter 15, 319–343.

Tinsley, T.W. and D. C. Kelly. 1970. An interim nomenclature system for the iridescent group of viruses. *Journal of Invertebrate Pathology* 16(3):470–472.

Todd, John F. J. 2005. Ion trap mass spectrometer–past, present, and future (?) *Mass Spectrometry Reviews* 10(1):3–52.

Tonella, L., C. Hoogland, P. A. Binz et al. 2001. New perspectives in the *Escherichia coli* proteome investigation. *Proteomics* 1(3):409–423.

Tonka, T. and J. Weiser. 2000. Iridovirus infection in mayfly larvae. *Journal of Invertebrate Pathology* 76(3):229–231.

Trainor, J. R. and P. J. Derrick. 1992. *Mass Spectrometry in the Biological Sciences: A Tutorial.* Dordrecht, the Netherlands: Kluwer Academic Publishers, pp. 3–27.

Turnbough, Jr., C. L. 2003. Discovery of phage display peptide ligands for species-specific detection of *Bacillus* spores. *Journal of Microbiological Methods* 53(2):263–271.

U.S. Environmental Protection Agency. 1999. *Method 1623: Cryptosporidium and Giardia in Water by Filtration/IMS/FA EPA-821-R-99-006.* Washington, DC: U.S. Environmental Protection Agency, Office of Water.

Vaidyanathan, S., D. B. Kell, and R. Goodacre. 2002. Flow-injection electrospray ionization mass spectrometry of crude cell extracts for high-throughput bacterial identification. *Journal of the American Society for Mass Spectrometry* 13(2):118–128.

Vaidyanathan, S., J. J. Rowland, D. B. Kell, and R. Goodacre. 2001. Discrimination of aerobic endospore-forming bacteria via electrospray-ionization mass spectrometry of whole cell suspensions. *Analytical Chemistry* 73(17):4134–4144.

Valaskovic, G. A., L. Utley, M. S. Lee, and J. T. Wu. 2006. Ultra-low flow nanospray for the normalization of conventional liquid chromatography/mass spectrometry through equimolar response: Standard-free quantitative estimation of metabolite levels in drug discovery. *Rapid Communications in Mass Spectrometry* 20(7):1087–1096.

Valentine, N., S. Wunschel, D. Wunschel et al. 2005. Effect of culture conditions on microorganism identification by matrix-assisted laser desorption ionization mass spectrometry. *Applied and Environmental Microbiology* 71(1):58–64.

Van Baar, B. L. 2000. Characterisation of bacteria by matrix-assisted laser desorption/ionisation and electrospray mass spectrometry. *FEMS Microbiology Reviews* 24(2):193–219.

Van Engelsdorp, D., J. D. Evans, C. Saegerman et al. 2009. Colony collapse disorder: A descriptive study. *PLoS ONE* 4(8):e6481.

Van Ert, M. N., S. A. Hofstadler, Y. Jiang et al. 2004. Mass spectrometry provides accurate characterization of two genetic marker types in *Bacillus anthracis*. *BioTechniques* 37(4):642–651.

Van Wuijckhuijse, A., C. Kientz, B. V. Baar et al. 2005. Development of Bioaerosol Alarming detector. In *Defence against Bioterror: Detection Technologies, Implementation Strategies and Commercial Opportunities*, D. Morrison, F. Milanovich, D. Ivnitski, and T. R. Austin, eds. Dordrecht, the Netherlands: Springer. Chapter 9, pp. 119–128.

Vanrobaeys, F., R. Van Costerk, G. Dhondt et al. 2005. Profiling of myelin proteins by 2D-gel electrophoresis and multidimensional liquid chromatography coupled to MALDI TOF-TOF mass spectrometry. *Journal of Proteome Research* 4(6):2283–2293.

VerBerkmoes, N. C., H. M. Connelly, C. Pan, and R. L. Hettich. 2004. Mass spectrometric approaches for characterizing bacterial proteomes. *Expert Review of Proteomics* 1(4): 433–447.

VerBerkmoes, N. C., W. J. Hervey, M. Shah et al. 2005. Evaluation of "Shotgun" proteomics for identification of biological threat agents in complex environmental matrices: Experimental simulations. *Analytical Chemistry* 77(3):923–932.

Verma, S. and K. P. S. Phogat. 1982. Seasonal incidence of *Apis* iridescent virus in *Apis cerana indica*. Uttar Pradesh, India. *Indian Bee Journal* 44:36–37.

Vestal, M. L., P. Juhasz, and S. A. Martin. 1995. Delayed extraction matrix-assisted time-of-flight mass spectrometry. *Rapid Communications in Mass Spectrometry* 9(11): 1044–1050.

Vollmer, M., E. Nagele, and P. J. Horth. 2003. Differential proteome analysis: Two-dimensional nano-lc/ms of *E. coli* proteome grown on different carbon sources. *Journal of Biomolecular Techniques* 14(2):128–135.

Von Wintzingerode, F., S. Böcker, C. Schlötelburg et al. 2002. Base-specific fragmentation of amplified 16S rRNA genes analyzed by mass spectrometry: A tool for rapid bacterial identification. *Proceedings of the National Academy of Sciences of the United States of America* 99(10):7039–7044.

Vorm, O. and P. Roepstorff. 1994. Peptide sequence information derived by partial acid hydrolysis and matrix-assisted laser desorption/ionization mass spectrometry. *Biological Mass Spectrometry*. 23(12):734–740.

Wahl, K. H., S. C. Wunschel, K. H. Jarman et al. 2002. Analysis of microbial mixtures by matrix-assisted laser desorption/ionization time-of-flight mass spectrometry. *Analytical Chemistry* 74(24):6191–6199.

Walker, J., A. J. Fox, V. Edwards-Jones, and D. B. J. Gordon. 2002. Intact cell mass spectrometry (ICMS) used to type methicillin-resistant *Staphylococcus aureus*: Media effects and inter-laboratory reproducibility. *Journal of Microbiological Methods* 48(2–3):117–126.

Wallman, L., S. Ekström, M. Magnusson et al. 2006. Robotic implementation of a microchip-based protein clean-up and enrichment system for MALDI-TOF MS readout. *Measurement Science and Technology* 17(12):3147–3153.

Wallman, L., S. Ekstrom, G. M. Varga, T. Laurell, and J. Nielsson. 2004. Autonomous protein sample processing on-chip using solid-phase microextraction, capillary force pumping, and microdispensing. *Electrophoresis* 25(21–22):3778–3787.

Wang, J., R. S. Houk, D. Dreessen, and D. R. Wiederin. 1999. Speciation of trace elements in proteins in human and bovine serum by size exclusion chromatography and inductively coupled plasma-mass spectrometry with a magnetic sector mass spectrometer. *Journal of Biological Inorganic Chemistry* 4(5):546–553.

Wang, T., Y. Zhang, W. Chen et al. 2002. Reconstructed protein arrays from 3D HPLC/tandem mass spectrometry and 2D gels: Complementary approaches to Porphyromonas gingivalis protein expression. *Analyst* 127(11):1450–1456.

Wang, W., Z. Liu, L. Ma et al. 1999. Electrospray ionization multiple-stage tandem mass spectrometric analysis of diglycosyldiacylglycerol glycolipids from the bacteria *Bacillus pumilus*. *Rapid Communications in Mass Spectrometry* 13(12):1189–1196.

Wang, Y. X., J. W. Cooper, C. S. Lee, and D. L. DeVoe. 2004. Efficient electrospray ionization from polymer microchannels using integrated hydrophobic membranes. *Lab on a Chip* 4:363–367.

Wang, Z., K. Dunlop, S. R. Long, and L. Li. 2002. Mass spectrometric methods for generation of protein mass database used for bacterial identification. *Analytical Chemistry* 74(13): 3174–3182.

Wang, Z., L. Russon, L. Li et al. 1998. Investigation of spectral reproducibility in direct analysis of bacteria proteins by matrix-assisted laser desorption/ionization time-of-flight mass spectrometry. *Rapid Communications in Mass Spectrometry* 12(8):456–464.

Warscheid, B. and C. Fenselau. 2003. Characterization of *Bacillus* spore species and their mixtures using post source decay with a curved-field reflectron. *Analytical Chemistry* 75(20):5618–5627.

Warscheid, B. and C. Fenselau. 2004. A targeted proteomics approach to the rapid identification of bacterial cell mixtures by matrix-assisted laser desorption/ionization time-of-flight mass spectrometry. *Proteomics* 4(10):2877–2892.

Warscheid, B., K. Jackson, C. Sutton, and C. Fenselau. 2003. MALDI analysis of *Bacilli* in spore mixtures by applying a quadrupole ion trap time-of-flight tandem mass spectrometer. *Analytical Chemistry* 75(20):5608–5617.

Washburn, M. P., D. Wolters, and J. R. Yates, III. 2001. Large-scale analysis of yeast proteome by multidimensional protein identification technology. *Nature Biotechnology* 19(3):242–247.

Watt, B. E., S. L. Morgan, and A. Fox. 1991. 2-Butenoic acid, a chemical marker for poly-β-hydroxybutyrate identified by pyrolysis-gas chromatography/mass spectrometry in analyses of whole microbial cells. *Journal of Analytical and Applied Pyrolysis* 19:237–249.

Webby, R and J. Kalmakoff. 1998. Sequence comparison of the major capsid protein gene from 18 diverse iridoviruses. *Archives of Virology* 143(10):1940–1966.

Westin, L., C. Miller, D. Vollmer et al. 2001. Antimicrobial resistance and bacterial identification utilizing a microelectronic chip array. *Journal of Clinical Microbiology* 39(3):1097–1104.

Wheeler, D. L., T. Barrett, D. A. Benson et al. 2003. Database resources of the National Center for Biotechnology. *Nucleic Acids Research* 31(1):28–33.

Wick, C. H., M. F. Stanford, R. Jabbour et al. 2009. Pandemic (H1N1) 2009 Cluster analysis: A preliminary assessment. *Nature Precedings*, September 16, 2009. http://precedings. nature.com/documents/3773/version/1/files/npre20093773–1.pdf (accessed July 1, 2012).

Wick, C. H., M. F. Stanford, A. W. Zulich et al. 2010. Iridescent virus and *Nosema ceranae* linked to honey bee Colony Collapse Disorder (CCD). Edgewood Chemical Biological Center Technical Report TR-814.

Wick, C. H., M. F. Stanford, A. W. Zulich et al. 2012. Method for detection and identification of cell types. *US Patent 8,244,581* filed September 30, 2009 and issued July 17, 2012.

Wickman, G., B. Johansson, J. Bahar-Gogani et al. 1998. Liquid ionization chambers for absorbed dose measurements in water at low dose rates and intermediate photon energies. *Medical Physics* 25(6):900–907.

Wilkes, J. G., F. Rafii, J. B. Sutherland et al. 2006. Pyrolysis mass spectrometry for distinguishing potential hoax materials from bioterror agents. *Rapid Communications in Mass Spectrometry* 20(16):2383–2386.

Wilkes, J. G., L. G. Rushing, J. F. Gagnon et al. 2005b. Rapid phenotypic characterization of Vibrio isolates by pyrolysis metastable atom bombardment mass spectrometry. *Antonie Van Leeuwenhoek* 88(2):151–161.

Wilkes, J. G., L. Rushing, R. Nayak et al. 2005a. Rapid phenotypic characterization of *Salmonella enterica* strains by pyrolysis metastable atom bombardment mass spectrometry with multivariate statistical and artificial neural network pattern recognition. *Journal of Microbiological Methods* 61(3):321–334.

Wilkins, M. R., K. L. Williams, R. D. Appel, and D. F. Hochstrasser, eds. 1997. *Proteome Research: New Frontiers in Functional Genomics*. New York: Springer, p. 13.

Williams, D. D., O. Benedek, C. L. Turnbough, Jr. 2003a. Species-specific peptide ligands for the detection of *Bacillus anthracis* spores. *Applied and Environmental Microbiology* 69(10):6288–6293.

Williams, T. 1998. Invertebrate iridescent viruses. In *The Insect Viruses*, L. K. Miller and L. A. Ball, eds. New York: Springer, pp. 31–68.

Williams, T. 2008. Natural invertebrate hosts of iridoviruses (Iridoviridae). *Neotropical Entomology* 37(6):615–632.

Williams, T., D. Andrzejewski, J. O. Lay, Jr. et al. 2003. Experimental factors affecting the quality and reproducibility of MALDI TOF mass spectra obtained from whole bacteria cells. *Journal of the American Society for Mass Spectrometry* 14(4):342–351.

Williams T. and J. S. Cory. 1994. Proposals for a new classification of iridescent virus. *Journal of General Virology* 75(6):1291–1301.

Wilm, M.S. and M. Mann. 1994. Error-tolerant identification of peptides in sequence databases by peptide sequence tags. *Analytical Chemistry* 66:4390–4399.

Wilm, M. S. and M. Mann. 1996. Analytical properties of the nanoelectrospray ion source. *Analytical Chemistry* 68(1):1–8.

Wilson, W. T. and D. M. Menapace. 1979. Disappearing disease of honey bees: A survey of the United States. *American Bee Journal* 119:118–119.

Wisniewski, J. R., N. Nagaraj, A. Zougman, and M. Mann. 2009a. Filter aided sample preparation (FASP) combines the advantages of in-gel and in-solution digestion. In *Proceedings of the 57th ASMS Conference in Mass Spectrometry and Allied topics*. Philadelphia, PA: ASMS, June 2009.

Wisniewski, J. R., A. Zougman, N. Nagaraj, and M. Mann. 2009b. Spin filter-based sample preparation for shotgun proteomics. *Reply Nature Methods* 6:359–362.

Woese, C. R., E. Stackebrandt, T. J. Macke, and G. E. Fox. 1985. A phylogenetic definition of the major eubacterial taxa. *Systematic and Applied Microbiology* 61(2):143–151.

Wolffs, P., R. Knutsson, B. Norling, and P. Radstrom. 2004. Rapid quantification of *Yersinia enterocolitica* in pork samples by a novel sample preparation method, flotation, prior to real-time PCR. *Journal of Clinical Microbiology* 42(3):1042–1047.

Wolters, D. A., M. P. Washburn, and J. R. Yates, III. 2001. An automated multidimensional protein identification technology for shotgun proteomics. *Analytical Chemistry* 73(23):5683–5690.

Wong, P. S., N. Srinivasan, N. Kasthurikrishnan et al. 1996. On-line monitoring of the photolysis of benzyl acetate and 3,5-dimethoxybenzyl acetate by membrane introduction mass spectrometry. *Journal of Organic Chemistry* 61(19):6627–6632.

World Health Organization. 2010. Pandemic (H1N1) 2009. http://www.who.int/csr/disease/swineflu/en/ (accessed April 6, 2009).

Wright, Jr., G. L., L. H. Cazares, S. M. Leung et al. 1999. Proteinchip® surface enhanced laser desorption/ionization (SELDI) mass spectrometry: A novel protein biochip technology for detection of prostate cancer biomarkers in complex protein mixtures. *Prostate Cancer Prostatic Diseases* 2:264–276.

Wu, C. C., M. J. MacCoss, K. E. Howell, and J. R. Yates, III. 2003. A method for the comprehensive proteomic analysis of membrane proteins. *Nature Biotechnology* 21(5): 532–538.

Wu, J. and S. A. McLuckey. 2003. Ion/ion reactions of multiply charged nucleic acid anions: Electron transfer, proton transfer, and ion attachment. *International Journal of Mass Spectrometry* 228(2–3):577–597.

Wuhrer, M. and A. M. Deelder. 2006. Matrix-assisted laser desorption/ionization in-source decay combined with tandem time-of-flight mass spectrometry of permethylated oligosaccharides: Targeted characterization of specific parts of the glycan structure. *Rapid Communications in Mass Spectrometry* 20(6):943–951.

Wunschel, D. S., K. F. Fox, A. Fox et al. 1997. Quantitative analysis of neutral and acidic sugars in whole bacterial cell hydrolysates using high-performance anion-exchange liquid chromatography-electrospray ionization tandem mass spectrometry. *Journal of Chromatography A* 776(2):205–219.

Wunschel, S. C., K. H. Jarman, C. E. Petersen et al. 2005. Bacterial analysis by MALDI-TOF mass spectrometry: An inter-laboratory comparison. *Journal of the American Society for Mass Spectrometry* 16(4):456–462.

Xiang, F., G. Anderson, T. Veenstra, M. Lipton, and R. Smith. 2000. Characterization of microorganisms and biomarker development from global ESI-MS-MS analyses of cell lysates. *Analytical Chemistry* 72(11):2475–2481.

Xie, J., Y. N. Miao, J. Shih, Y. C. Tai, and T. D. Lee. 2005. Microfluidic platform for liquid chromatography-tandem mass spectrometry analyses of complex peptide mixtures. *Analytical Chemistry* 77(21):6947–6953.

Yanes, O., J. Villanueva, E. Querol, and F. X. Aviles. 2007. Detection of non-covalent protein interactions by 'intensity fading' MALDI-TOF mass spectrometry: Applications to proteases and protease inhibitors. *Nature Protocols* 2:119–130.

Yang, M., P. T. A. Reilly, K. B. Boraas, W. B. Whitten, and J. M. Ramsey. 1996. Real-time chemical analysis of aerosol particles using an ion trap mass spectrometer. *Rapid Communications in Mass Spectrometry* 10:347–351.

Yates, J. R., III and J. K. Eng. 2000. Identification of nucleotides, amino acids, or carbohydrates by mass spectrometry. *U.S. Patent 6,017,693*, issued January 25, 2000.

Yeung, S.-W., T. M.-H. Lee, H. Cai, and I.-M. Hsing. 2006. A DNA biochip for on-the-spot multiplexed pathogen identification. *Nucleic Acids Research* 34(18):118–125.

Yin, H., K. Killeen, R. Brennen et al. 2005. Microfluidic chip for peptide analysis with an integrated HPLC column, sample enrichment column, and nano-electrospray tip. *Analytical Chemistry* 77(2):527–533.

Yost, R. A. and C. G. Enke. 1979. TQMS for mixture analysis. *Analytical Chemistry* 51:1251A–1264A.

Yost, R. A. and C. G. Enke. 1983. TQMS. In *Tandem Mass Spectrometry*, F. W. McLafferty, ed. New York: Wiley, pp. 175–196.

Zambonin, C. G., C. D. Calvano, L. D'Accolti, and F. Palmisano. 2006. Laser desorption/ionization time-of-flight mass spectrometry of squalene in oil samples. *Rapid Communications in Mass Spectrometry* 20(2):325–327.

Zerega, Y., J. A. G. Brincourt, and R. Catella. 1994. A new operating mode of a quadrupole ion trap in mass spectrometry: Part 1. Signal visibility. *International Journal of Mass Spectrometry and Ion Processes* 132(1–2):57–72.

Zhang, H., M. Stoeckli, P. E. Andren, and R. M. Caprioli. 1999. Combining solid-phase preconcentration, capillary electrophoresis and off-line matrix-assisted laser desorption/ionization mass spectrometry: Intracerebral metabolic processing of peptide E in vivo. *Journal of Mass Spectrometry* 34(4):377–383.

Zhang, J., K. Schubothe, B. Li et al. 2005. Infrared multiphoton dissociation of O-linked mucin-type oligosaccharides. *Analytical Chemistry* 77(1):208–214.

Zhong, H. and L. Li. 2005. An algorithm for interpretation of low-energy collision-induced dissociation product ion spectra for de novo sequencing of peptides. *Rapid Communications in Mass Spectrometry* 19(8):1084–1096.

Zhou, X., G. Gonnet, M. Hallett et al. 2001. Cell fingerprinting: An approach to classifying cells according to mass profiles of digests of protein extracts. *Proteomics* 1:683–690.

Zubarev, R. A. 2004. Electron-capture dissociation tandem mass spectrometry. *Current Opinion in Biotechnology* 15(1):12–16.

Zubarev, R. A., N. L. Kelleher, and F. W. McLafferty. 1998. Electron capture dissociation of multiply charged protein cations: A nonergodic process. *Journal of the American Chemical Society* 120(13):3265–3266.

Index